Advanced Semiconductor Materials and Films: Properties and Applications

Advanced Semiconductor Materials and Films: Properties and Applications

Guest Editors

Pengfei Guo
Shaoding Liu
Chaoping Liu

MDPI

Basel • Beijing • Wuhan • Barcelona • Belgrade • Novi Sad • Cluj • Manchester

Guest Editors
Pengfei Guo
College of Physics and Optoelectronics
Taiyuan University of Technology
Taiyuan
China

Shaoding Liu
College of Physics and Optoelectronics
Taiyuan University of Technology
Taiyuan
China

Chaoping Liu
Department of Physics
Shantou University
Shantou
China

Editorial Office
MDPI AG
Grosspeteranlage 5
4052 Basel, Switzerland

This is a reprint of the Special Issue, published open access by the journal *Coatings* (ISSN 2079-6412), freely accessible at: https://www.mdpi.com/journal/coatings/special_issues/Semiconductor_Films.

For citation purposes, cite each article independently as indicated on the article page online and as indicated below:

Lastname, A.A.; Lastname, B.B. Article Title. *Journal Name* **Year**, *Volume Number*, Page Range.

ISBN 978-3-7258-7350-0 (Hbk)
ISBN 978-3-7258-7351-7 (PDF)
https://doi.org/10.3390/books978-3-7258-7351-7

Contents

About the Editors

Pengfei Guo

Pengfei Guo received his Master and Doctoral Degree from the College of Materials Science and Engineering of Hunan University in 2014. He was selected in the "Hong Kong Scholars Program" in 2015, then worked as a Postdoctoral Research Fellow and Hong Kong Scholar in the Department of Physics in City University of Hong Kong from 2016 to 2018. He has authored over 70 SCI publications, including 40 papers as first or corresponding author in prestigious journals such as *Advanced Materials, Physical Review Letters, ACS Nano, Nano Letters, Advanced Functional Materials, Nano Energy* et al. His research contributions have garnered over 3000 citations in leading journals. Now, he is a full professor in College of Physics and Optoelectronics in Taiyuan University of Technology. His research interests include low-dimensional semiconductor materials, optoelectronic, nanophotonics, and energy technologies.

Shaoding Liu

Shaoding Liu is affiliated with Taiyuan University of Technology. His research interests include micro- and nano-scale nonlinear optics and plasmonics. He has presided over more than ten funded projects, including two NSFC General Programs, one NSFC Youth Program, the Shanxi Outstanding Youth Fund, and a key research and development plan. He has published over 80 papers in academic journals, which have received total citations exceeding 2000, and has applied for/been granted more than 10 national invention patents. His achievements include being recognized as an Outstanding Young Academic Leader in Shanxi Higher Education Institutions, an Outstanding Mid-Career Innovative Talent of the Shanxi "131" Project, a Shanxi Youth "Sanjin Scholar", and a Shanxi Youth Top-Notch Talent. He also received the Second Prize of the Shanxi Natural Science Award as the first accomplished person.

Chaoping Liu

Chaoping Liu is a faculty member in the Department of Physics at Shantou University, where he focuses his research on the material physics and device applications of wide-bandgap oxides and copper-based halide semiconductors. He has authored over 70 SCI publications, including 30 papers as first or corresponding author in prestigious journals such as *Nature Communications* (featured as a Highlight), *Physical Review Applied et al.* His research contributions have garnered over 2,300 citations in leading journals like *Nature Materials* and are supported by two authorized Chinese invention patents. Beyond his experimental work, Liu leads several high-level research initiatives, serving as the principal investigator for the General Program of the National Natural Science Foundation of China (NSFC) and the Guangdong Provincial International Science and Technology Cooperation Project.

coatings

Editorial

Progress and Challenges of Bandgap Engineering in One-Dimensional Semiconductor Materials

Xia Shen [1] and Pengfei Guo [2,*]

1 School of Basics, Shanxi Institute of Energy, Taiyuan 030600, China
2 College of Physics and Optoelectronics, Taiyuan University of Technology, Taiyuan 030024, China
* Correspondence: guopengfei@tyut.edu.cn

Since their emergence, low-dimensional nanostructures, particularly semiconductor nanowires and nanobelts, have attracted significant attention due to their unique properties and potential applications in future integrated optoelectronic devices and circuits. With the continuous development of semiconductor technology, the need for miniaturization, high integration, and low energy consumption in devices has led to higher requirements for nanomaterials [1]. Semiconductor material-based integration and the diversification of devices pose major challenges. Thus, the ability to modulate the bandgap along a single nanostructure greatly enhances functionality in optoelectronics. Therefore, studies in this area are essential in order to pave the way for highly integrated future devices.

On the other hand, low-dimensional nanostructures have been reported for use in various optoelectronic devices, including nanoscale lasers, LED, solar cells, waveguides, sensors, photodetectors, spectrometers, and solid-state light sources [2]. However, there are many challenges in their practical application that restrict further development and large-scale application. This Special Issue highlights key areas where significant challenges remain, for example, the structural evaluation and stability of organic semiconductors [3], the doping and stability of inorganic perovskites [4], the external quantum efficiency of light-emitting diodes [5], morphological engineering and semiconductor coupling in photocatalysis [6], and the effects of ions doping in semiconductors [7]. Currently, considering the shortage of available materials and the performance improvement bottleneck, contradictions are particularly prominent [8]. For example, the exponential growth of data in the information age demands enhanced security and privacy protection for image transmission. Photodetectors that convert light signals into electrical signals play a pivotal role in multi-color imaging, optical communications, and logical operations [9]. Consequently, developing high-performance multifunctional photodetectors is essential in order to meet the future demands of information technology. Although relevant research has been reported, achieving multifunctionality and high safety in devices remains a major challenge, highlighting the importance of manufacturing new types of equipment [10]. Interestingly, the gradient structure, with its funnel-shaped band structure, enhances the transport and transmission of carriers, providing a foundation for the development of multifunctional platform optoelectronic devices [11].

So far, significant research efforts have been devoted by scientists around the world to achieving bandgap engineering along low-dimensional semiconductor structures because of their ideal properties and potential applications in photoelectronic circuits [12]. For example, $Ga_xIn_{1-x}As$-InAs heterostructure arrays were fabricated via the molecular beam epitaxy (MBE) approach through hetero-interface diffusion [13]. Si-Ge nanowires with abrupt hetero-interfaces [14] and Si-SiGe nanowire superlattices [15] were developed via the

 https://doi.org/10.3390/coatings16010036

chemical vapor deposition (CVD) approach. GaP-GaAs superlattice nanostructures were prepared using the laser-assisted CVD method, indicating atomic abrupt interfaces along the junctions [16]. InGaN-GaN quantum-well nanowires were synthesized via the metal–organic CVD (MOCVD) approach [17], demonstrating a room-temperature nanoscale laser application. CdS-SnS_2 microwire superlattices were synthesized via the CVD approach and showed weak optical nonlinearities and slow light engineering [18]. $CsPbX_3$ (X = Cl, Br, I) nanowires were synthesized via an anion exchange process, which showed a white (R-G-B) emission along a single nanowire [19]. In addition, $CsPbCl_3$-$CsPbI_3$ heterostructure nanowires were fabricated through a CVD approach and could act as dual-wavelength lasers [20]. Nevertheless, bandgap engineering in a single nanowire has always posed a significant challenge for the scientific community.

Scientists are exploring preparation strategies for nanomaterials and are expected to continuously improve on characteristic methods such as MBE, CVD, MOCVD, MOVPE, laser-assisted CVD, and the ion exchange method. CVD is a useful and reliable technique for synthesizing various one-dimensional nanomaterials with high crystal quality due to its advantages of high controllability, flexible scalability, high yields, and low cost [21]. However, achieving multifunctionality and strong security within a single device is a significant challenge. Thus, achieving compositional modulation along the nanostructure represents a useful strategy in the development of multifunctional devices and circuits.

Based on this consideration, in 2011, the magnetic-pulling, source-moving CVD approach was developed to synthesize numerous compositional gradient nanostructures, for example, $CdS_{1-x}Se_x$ nanowires [22], CdS/$CdS_{1-x}Se_x$ heterostructure nanoribbons [9], and ZnCdSSe bandgap gradient nanoribbons [23]. They offer several advantages compared with composition homogenous structures. Firstly, the magnetic-pulling, source-moving CVD method represents an improvement on the traditional one. This method can fabricate bandgap gradient nanowires (e.g., $CdS_{1-x}Se_x$ nanowires and ZnCdSSe nanowires), which exhibit continuously tunable band gaps ranging from blue to green and then to red [24]. Secondly, photodetectors based on these bandgap gradient nanowires exhibit superior photoelectric performance, such as high responsivity, high detectivity, and fast response speed in the millisecond magnitude, which are superior to traditional devices [25,26]. Thirdly, the synthesis of high-quality on-wire compositional gradient nanostructures with high-performance photoconductance provides an ideal material platform for investigating gradient structural characteristics more clearly and promoting the development of all-optical logic gates, solid-state white light sources, and encrypted communications.

We therefore believe that on-nanowire bandgap modulation has many potential applications in the field of nanoscience and nanotechnology and will make a valuable contribution to the nanoworld in the future.

Author Contributions: Conceptualization, X.S. and P.G.; methodology, X.S. and P.G.; formal analysis, X.S.; investigation, X.S.; resources, X.S. and P.G.; writing—original draft preparation, X.S. and P.G.; writing—review and editing, P.G.; visualization, P.G.; supervision, P.G. All authors have read and agreed to the published version of the manuscript.

Acknowledgments: The authors are grateful to the National Natural Science Foundation of China (No. 52373246) for financial support.

Conflicts of Interest: The authors declare no conflicts of interest.

References

1. Yang, Z.Y.; Albrow, T.; Cui, H.X.; Alexander, W.J.; Gu, F.X.; Wang, X.M.; Wu, T.C.; Zhu, M.H.; Williams, C.; Wang, P.; et al. Single-nanowire spectrometers. *Science* **2019**, *365*, 101. [CrossRef] [PubMed]
2. Quan, L.; Kang, J.; Ning, C.; Yang, P. Nanowires for photonics. *Chem. Rev.* **2019**, *119*, 9153. [CrossRef]

https://doi.org/10.3390/coatings16010036

3. Noreen, S.; Aljaafreh, M.J.; Sumrra, S.H. Modeling Enthalpy of Formation with Machine Learning for Structural Evaluation and Thermodynamic Stability of Organic Semiconductors. *Coatings* **2025**, *15*, 758. [CrossRef]
4. Lu, X.; Wu, H.; Xu, J.; Chen, J.; Huang, Y.; Li, H.; Song, J.; Huang, R. Influence of Ce3+ Doping on Photoluminescence Properties and Stability of Cs4SnBr6 Zero-Dimensional Perovskite. *Coatings* **2024**, *14*, 945. [CrossRef]
5. Li, X.; Shen, X.; Lv, Q.; Guo, P.; Xiao, L. Metal Halide Perovskites: Promising Materials for Light-Emitting Diodes. *Coatings* **2024**, *14*, 83. [CrossRef]
6. Xu, F.; Zhang, K.; Li, K.; Ju, H.; Xue, Q.; Qi, X.; Jiang, J. The Construction of an α-F2O3/Tubular g-C3N4 Z-Scheme Heterojunction Catalyst for the Efficient Photocatalytic Degradation of Tetracycline. *Coatings* **2023**, *13*, 1909. [CrossRef]
7. Sun, C.; Lou, Z.; Ai, X.; Xue, Z.; Zhang, H.; Chen, G. Effects of Nitrogen Doping on Pulling Rate Range of Defect-Free Crystal in CZ Silicon. *Coatings* **2023**, *13*, 1637. [CrossRef]
8. Ning, C.; Dou, L.; Yang, P. Bandgap engineering in semiconductor alloy nanomaterials with widely tunable compositions. *Nat. Rev. Mater.* **2017**, *2*, 17070. [CrossRef]
9. Perea, D.E.; Li, N.; Dickerson, R.M.; Misra, A.; Picraux, S.T. Controlling heterojunction abruptness in VLS-grown semiconductor nanowires via in situ catalyst alloying. *Nano Lett.* **2011**, *11*, 3117. [CrossRef]
10. Kim, C.; Lee, H.; Cho, Y.; Yang, J.; Lee, R.; Lee, J.; Jo, M. On-nanowire band-graded Si:Ge photodetectors. *Adv. Mater.* **2011**, *23*, 1025. [CrossRef]
11. Chen, P.; Zhang, Z.W.; Duan, X.D.; Duan, X.F. Chemical synthesis of two-dimensional atomic crystals, heterostructures and superlattices. *Chem. Soc. Rev.* **2018**, *47*, 3129. [CrossRef]
12. Ren, D.; Ahtapodov, L.; Nilsen, J.; Yang, J.; Gustafsson, A.; Huh, J.; Conibeer, G.; Helvoort, A.; Fimland, B.; Weman, H. Single-mode near-infrared lasing in a GaAsSb-based nanowire superlattice at room temperature. *Nano Lett.* **2018**, *18*, 2304. [CrossRef] [PubMed]
13. Krogstrup, P.; Yamasaki, J.; Sørensen, C.; Johnson, E.; Wagner, J.; Pennington, R.; Aagesen, M. Junctions in axial III−V heterostructure nanowires obtained via an interchange of group III elements. *Nano Lett.* **2009**, *9*, 3689. [CrossRef]
14. Clark, T.; Nimmatoori, P.; Lew, K.; Pan, L.; Redwing, J.; Dickey, E. Diameter dependent growth rate and interfacial abruptness in vapor–liquid–solid Si/Si1−xGex heterostructure nanowires. *Nano Lett.* **2008**, *8*, 1246. [CrossRef] [PubMed]
15. Wu, Y.; Fan, R.; Yang, P. Block-by-block growth of single-crystalline Si/SiGe superlattice nanowires. *Nano Lett.* **2002**, *2*, 83. [CrossRef]
16. Gudiksen, M.; Lauhon, L.; Wang, J.; Smith, D.; Lieber, C. Growth of nanowire superlattice structures for nanoscale photonics and electronics. *Nature* **2002**, *415*, 617. [CrossRef]
17. Qian, F.; Li, Y.; Gradečak, S.; Park, H.; Dong, Y.; Ding, Y.; Lieber, C. Multi-quantum-well nanowire heterostructures for wavelength-controlled lasers. *Nat. Mater.* **2008**, *7*, 701. [CrossRef]
18. Dai, G.; Zou, B.; Wang, Z. Preparation and periodic emission of superlattice CdS/CdS: SnS2 microwires. *J. Am. Chem. Soc.* **2010**, *132*, 12174. [CrossRef]
19. Dou, L.; Lai, M.; Kley, C.; Yang, Y.; Bischak, C.; Zhang, D.; Eaton, S.; Ginsberg, N.; Yang, P. Spatially resolved multicolor CsPbX3 nanowire heterojunctions via anion exchange. *Proc. Natl. Acad. Sci. USA* **2017**, *114*, 7216. [CrossRef]
20. Guo, P.; Liu, D.; Shen, X.; Lv, Q.; Wu, Y.; Yang, Q.; Yu, K. On-wire axial perovskite heterostructures for monolithic dual-wavelength laser. *Nano Energy* **2021**, *92*, 106778. [CrossRef]
21. Fan, F.; Turkdogan, S.; Liu, Z.; Shelhammer, D.; Ning, C. A monolithic white laser. *Nat. Nanotechnol.* **2015**, *10*, 796. [CrossRef]
22. Gu, F.; Yang, Z.; Yu, H.; Xu, J.; Wang, P.; Tong, L.; Pan, A. Spatial bandgap engineering along single alloy nanowires. *J. Am. Chem. Soc.* **2011**, *133*, 2037. [CrossRef] [PubMed]
23. Yang, Z.; Xu, J.; Wang, P.; Zhuang, X.; Pan, A.; Tong, L. On-nanowire spatial bandgap design for white light emission. *Nano Lett.* **2011**, *11*, 5085. [CrossRef] [PubMed]
24. Guo, P.; Yang, Q.; Shen, X.; Lv, Q.; Hao, Y.; Xiao, L.; Ho, J.; Yu, K. Room-Temperature Broad-Wavelength-Tunable Single-Mode Lasing from Alloyed $CdS_{1-x}Se_x$ Nanotripods. *ACS Nano* **2022**, *16*, 12767. [CrossRef]
25. Xu, Z.; Lv, Q.; Li, X.; Meng, Y.; Ho, J.; Guo, P. Composition-Tunable Bandgap Engineering of Horizontally Guided CdS_xSe_{1-x} Nanowalls for High-Performance Photodetectors. *ACS Appl. Mater. Interfaces* **2025**, *17*, 1962. [CrossRef] [PubMed]
26. Guo, P.; Wang, J.; Shen, X.; Lv, Q.; Li, X.; Xu, Z.; Han, S.; Bian, Y.; Meng, Y.; Yang, L.; et al. Bandgap engineering of halide perovskite nanoribbons for highperformance photodetection. *Nano Res.* **2025**, *18*, 94907347. [CrossRef]

https://doi.org/10.3390/coatings16010036

coatings

Editorial

Special Issue: Advanced Semiconductor Materials and Films: Properties and Applications

Xia Shen, Qihang Lv and Pengfei Guo *

College of Physics and Optoelectronics, Key Laboratory of Advanced Transducers and Intelligent Control System Ministry of Education, Taiyuan University of Technology, Taiyuan 030024, China
* Correspondence: guopengfei@tyut.edu.cn

Advanced semiconductor materials and films are building blocks for multifunctional devices and circuits, integrated optoelectronic chips, and high-throughput communications, which have proved basic material platforms for nanoscience and technologies. With unique structural properties and having potential to act as significant optoelectronic and nanophotonic units, advanced semiconductor materials, e.g., quantum dots, one-dimensional (1D) nanostructures, 2D materials, and semiconductor thin films, have been used in a variety of fields. In particular, 1D bandgap modulated structures, heterostructures, and artificial optical systems (fabricated via a chemical vapor deposition (CVD), metal-organic chemical vapor deposition (MOCVD), molecular beam epitaxy (MBE), atomic layer deposition (ALD), solution-based process, and sputtering, etc., may provide new opportunities for wavelength-tunable lasers, photovoltaics, solid-state lighting, spectrometers, and multicolor displays, as well as other highly integrated optoelectronic devices.

Over the past decade, tremendous progress has been made in the development of advanced semiconductor materials. This Special Issue aims to provide a forum for researchers to share current research findings and investigations to promote further research into the highly integrated photonic and optoelectronic performance of advanced multifunctional materials, including experimental and theoretical calculations. Authors are invited to submit original research articles, critical review articles, or short communications focused on, but not limited to, the following topics: the on-structure synthesis of 1D or 2D materials with radially or axially modulated compositions along a single nanostructure; innovative techniques for the realization of bandgap engineering on a single nanowire or 2D materials; recent advances in thin films and optoelectronic applications; the fabrication and physical properties of semiconductor materials; and semiconductor–metal interactions or light–mater interactions, etc.

Most of the traditional semiconductor materials have too large a footprint to be suitable for applications in the miniaturization of on-chip photonic circuits and systems. Thus, a composition tunable within a single nanostructure is urgently needed for low-power consumption, miniaturization, and high-integration devices [1]. Due to versatile physical/chemical properties, semiconductor nanostructures with axial or lateral on-wire bandgap engineering have emerged as building blocks for fundamental nanophotonic and optoelectronic devices [2,3]. Prof. Pan and Prof. Tong reported the created of axial bandgap-graded CdSSe nanowires and ZnCdSSe composition gradient nanowires through a magnetic-pulled CVD method, which were used as solid-state light sources, wavelength tunable lasers, and photodetectors [1]. Prof. Jo and Prof. Picraux reported the creation of composition-graded Si_xGe_{1-x} (x = 0–1) nanowires by means of an in situ catalyst-alloyed CVD approach and a thermal evaporation process [4,5]. Additionally, this magnetic-pulled system was also applied in fabricating 2D composition-graded nanoribbons as well as heterostructure nanoribbons. Prof. Ning reported lateral ZnCdSSe heterostructure nanoribbons, which were used to fabricate monolithic white lasers [6]. Composition-graded CdSSe alloy nanoribbons and CdSSe/CdS heterostructure nanoribbons were reported in

Prof. Pan's group, which were used to realize nanoscale dual-color and multi-color lasing [7,8]. Besides traditional semiconductor materials, novel perovskite nanomaterials are being developed rapidly in the application of nanophotonic and optoelectronic materials. For example, spatially resolved multi-component $CsPbX_3$ (X = Cl, Br, I) nanowires and $CsPbCl_3$-$CsPbI_3$ axial perovskite heterostructure nanowires were constructed to realize dual-wavelength lasing and asymmetric waveguide [9,10].

In recent years, transition metal dichalcogenides (TMDs), such as WS_2, MoS_2, WSe_2, $MoSe_2$ and van der Waals (vdWs) heterostructures with a nearly perfect crystalline structure and dangling-bond-free surface, have attracted considerable interest, making TMDs materials useful for flexible optoelectronic applications [11]. For example, Prof. Duan reported the synthesis of $WS_{2x}Se_{2-2x}$ alloy nanosheets with composition-tunable electronic properties [12]. A general synthetic strategy for 2D vdWs heterostructure arrays, including VSe_2/WSe_2, VSe_2/MoS_2, $NiTe_2/WSe_2$, $NbTe_2/WSe_2$ etc., was successfully realized in the same group, which acted as high-performance transistors [13]. Moreover, Zhao et al. reported a straightforward approach to realizing high-order SnS_2/WSe_2 vdWs superlattices by rolling up vdWs heterostructures [14]. This strategy can be extended to create diverse 2D/2D vdWs superlattices, more complex 2D/2D/2D vdWs superlattices, and beyond-2D materials.

Although 1D semiconductor nanostructures and 2D materials are of particular interest with respect to their unique low-dimensional footprint and versatile physical properties, several research groups have shown critical insights into disordered nanomaterials for integrated optoelectronic devices and systems. Prof. Joselevich reported a large-scale guided growth of horizontal nanowire arrays on sapphire substrates, including GaN nanowires, CdS nanowalls, $CsPbBr_3$ nanowires and CdSe/CdS nanowire heterostructures, which have devoted significant efforts to the development of high integrated devices and circuits in optoelectronic and nanotechnology [15]. The authors claimed that the guided growth of horizontal nanowires demonstrated in their group could be extended to a large variety of semiconductor materials, which were suitable for a wide range of applications in the future [15,16]. Prof. Li reported the fabrication of a wide variety of InP nanostructures with smooth sidewalls using an inverse metal-assisted chemical etching method without UV irradiation, which may find applications in transistors and optoelectronic devices with better performance and at lower cost than conventional etching methods [17].

In summary, advanced semiconductor materials provide new opportunities for integrated nanophotonics and optoelectronics. The results from recent investigations suggest that low-dimensional semiconductor nanostructures and thin films are excellent candidates for detectors, nanolasers, solid-state light sources, transistors, light emitting diodes, spectrometers, etc. More research efforts will be inspired to address the challenges in the optoelectronics field in the future.

Author Contributions: Conceptualization, X.S. and P.G.; investigation, X.S. and Q.L.; writing—original draft preparation, P.G. and X.S.; writing—review and editing, P.G.; supervision, P.G.; funding acquisition, P.G. All authors have read and agreed to the published version of the manuscript.

Funding: The authors are grateful to the start-up funding from Taiyuan University of Technology and Shanxi Basic Research Program Project (No. 20210302123128) for financial support.

Conflicts of Interest: The authors declare no competing financial interest.

References

1. Ning, C.Z.; Dou, L.; Yang, P. Bandgap engineering in semiconductor alloy nanomaterials with widely tunable compositions. *Nat. Rev. Mater.* **2017**, *2*, 17070. [CrossRef]
2. Quan, L.N.; Kang, J.; Ning, C.Z.; Yang, P.D. Nanowires for photonics. *Chem. Rev.* **2019**, *119*, 9153–9169. [CrossRef] [PubMed]
3. Yang, Z.Y.; Albrow, T.; Cui, H.X.; Alexander, W.J.; Gu, F.X.; Wang, X.M.; Wu, T.C.; Zhu, M.H.; Williams, C.; Wang, P.; et al. Single-nanowire spectrometers. *Science* **2019**, *365*, 1017–1020. [CrossRef] [PubMed]
4. Yang, J.E.; Park, W.H.; Kim, C.J.; Kim, Z.H.; Jo, M.H. Axially graded heteroepitaxy and Raman spectroscopic characterizations of $Si_{1-x}Ge_x$ nanowires. *Appl. Phys. Lett.* **2008**, *92*, 263111. [CrossRef]

5. Perea, D.E.; Li, N.; Dickerson, R.M.; Misra, A.; Picraux, S.T. Controlling heterojunction abruptness in VLS-grown semiconductor nanowires via in situ catalyst alloying. *Nano Lett.* **2011**, *11*, 3117–3122. [CrossRef] [PubMed]
6. Fan, F.; Turkdogan, S.; Liu, Z.C.; Shelhammer, D.; Ning, C.Z. A monolithic white laser. *Nat. Nanotech.* **2015**, *10*, 796–803. [CrossRef] [PubMed]
7. Zhuang, X.; Guo, P.; Zhang, Q.; Liu, H.; Li, D.; Hu, W.; Zhu, X.; Zhou, H.; Pan, A. Lateral composition-graded semiconductor nanoribbons for multi-color nanolasers. *Nano Res.* **2016**, *9*, 933–941. [CrossRef]
8. Xu, J.; Ma, L.; Guo, P.; Zhuang, X.; Zhu, X.; Hu, W.; Duan, X.; Pan, A. Room-temperature dual-wavelength lasing from single nanoribbon lateral heterostructures. *J. Am. Chem. Soc.* **2012**, *134*, 12394–12397. [CrossRef] [PubMed]
9. Dou, L.; Lai, M.; Kley, C.S.; Yang, Y.; Bischak, C.G.; Yang, P.D. Spatially resolved multicolor $CsPbX_3$ nanowire heterojunctions via anion exchange. *Proc. Natl. Acad. Sci. USA* **2017**, *114*, 7216–7221. [CrossRef] [PubMed]
10. Guo, P.F.; Liu, D.; Shen, X.; Lv, Q.H.; Wu, Y.; Yang, Q.; Li, P.; Hao, Y.; Ho, J.C.; Yu, K.M. On-wire axial perovskite heterostructures for monolithic dual-wavelength laser. *Nano Energy* **2021**, *92*, 106778. [CrossRef]
11. Chen, P.; Zhang, Z.W.; Duan, X.D.; Duan, X.F. Chemical synthesis of two-dimensional atomic crystals, heterostructures and superlattices. *Chem. Soc. Rev.* **2018**, *47*, 3129–3151. [CrossRef]
12. Duan, X.; Wang, C.; Fan, Z.; Hao, G.; Kou, L.; Halim, U.; Li, H.; Wu, X.; Wang, Y.; Jiang, J.; et al. Synthesis of $WS_{2x}Se_{2-2x}$ alloy nanosheets with composition-tunable electronic properties. *Nano Lett.* **2016**, *16*, 264–269. [CrossRef]
13. Li, J.; Yang, X.; Liu, Y.; Huang, B.; Wu, R.; Zhang, Z.; Zhao, B.; Ma, H.; Dang, W.; Wei, Z.; et al. General synthesis of two-dimensional van der Waals heterostructure arrays. *Nature* **2020**, *579*, 368–374. [CrossRef] [PubMed]
14. Zhao, B.; Wan, Z.; Liu, Y.; Xu, J.; Yang, X.; Shen, D.; Zhang, Z.; Guo, C.; Qian, Q.; Li, J.; et al. High-order superlattices by rolling up van der Waals heterostructures. *Nature* **2021**, *591*, 385–390. [CrossRef] [PubMed]
15. Tsivion, D.; Schvartzman, M.; Popovitz-Biro, R.; Huth, P.; Joselevich, E. Guided growth of millimeter-long horizontal nanowires with controlled orientations. *Science* **2011**, *333*, 1003. [CrossRef] [PubMed]
16. Schvartzman, M.; Tsivion, D.; Mahalu, D.; Raslin, O.; Joselevich, E. Self-integration of nanowires into circuits via guided growth. *Proc. Natl. Acad. Sci. USA* **2013**, *110*, 15195. [CrossRef] [PubMed]
17. Kim, S.H.; Mohseni, P.K.; Song, Y.; Ishihara, T.; Li, X.L. Inverse metal-assisted chemical etching produces smooth high aspect ratio InP nanostructures. *Nano Lett.* **2015**, *15*, 641–648. [CrossRef] [PubMed]

coatings

Review

Metal Halide Perovskites: Promising Materials for Light-Emitting Diodes

Yuyang Li [1], Xia Shen [1,*], Qihang Lv [1], Pengfei Guo [1,2,*] and Liantuan Xiao [1]

1 College of Electronic Information and Optical Engineering, Taiyuan University of Technology, Taiyuan 030024, China; 17390928994@163.com (X.L.); lv_qihang@163.com (Q.L.); xlt@sxu.edu.cn (L.X.)
2 Department of Materials Science and Engineering, City University of Hong Kong, Kowloon, Hong Kong 999077, China
* Correspondence: shenxia2019@126.com (X.S.); guopengfei2010@126.com (P.G.)

Abstract: Metal halide perovskites have shown excellent optoelectronic properties, including high photoluminescence quantum yield, tunable emission wavelengths, narrow full-width at half-maximums and a low-cost, solution-processed fabrication, which make it exhibit great potential as emission-layer materials of light-emitting diodes. With the joint efforts of researchers from different disciplines, there has been a significant progress in the improvement in the external quantum efficiency (EQE) and stability of perovskite light-emitting diodes (PeLEDs) in recent years, especially in green PeLEDs with EQEs over 30%. However, their operational stability lags behind other commercial organic and chalcogenide quantum dot emitters, limiting their practical application. In this review, we first introduce the basic device structure of PeLEDs, as well as the factors influencing the EQE and stability of PeLEDs. Secondly, the development of lead-based and lead-free PeLEDs are summarized systematically. Thirdly, challenges of PeLEDs are discussed in detail, including low the EQE of blue PeLEDs, poor device stability and EQE roll-off. Finally, some suggestions and perspectives for future research directions for PeLEDs are proposed.

Keywords: metal halide perovskites; light-emitting diodes; external quantum efficiency; stability; photonics

1. Introduction

Metal halide perovskites refer to a class of materials with a generic formula ABX_3, and their typical structure is constructed by corner-sharing metal halide $[BX_6]^{4-}$ octahedra and a monovalent cation (A) occupying the central vacancy between the octahedra, where A is Cs^+, MA^+ or FA^+; B is commonly Pb^{2+} or Sn^{2+}; and X is Cl^{-1}, Br^{-1} or I^{-1}. These materials not only have the benefits of high photoluminescence quantum yields (PLQYs), tunable bandgaps and high color purity, but they can also be fabricated easily via a sample chemical vapor deposition (CVD) method or solution-processing [1–3]. These merits enable the metal halide perovskite as the excellent emission layer in light-emitting diodes to satisfy the new International Telecommunication Union Rec. 2020 display standard [4–6]. Since the impressive work of perovskite light-emitting diodes (PeLEDs) with a low EQE of 0.76% reported by Tan et al. in 2014 [7], PeLEDs have achieved prosperous development with impressive peak EQE values for blue, green, red and near-infrared PeLEDs of 18.65%, 30.84%, 25.8% and 23.8%, respectively [8–11], making them fierce competitors for current commercial light emitters, such as organic LEDs and chalcogenide quantum dot LEDs (Table 1). The common device structure of PeLEDs is shown in Figure 1b, which includes the substrate, charge injection layers, charge blocking layers, perovskite emission layer and electrode (Figure 2a) [12]. When charges are injected from charge injection layers into the bulk perovskite emission layer, the generated free carriers are easily trapped by defects; with the further increase in current density, traps become saturated, so the free carriers can recombine radiatively to emit photons. As for quasi-2D and 0D perovskites, the injected charges tend to form excitons due to the quantum confinement effect. The excitons will rapidly undergo radiative recombination in 0D perovskites. Moreover, there are extra charges or energy cascades from low- to high-<n> phases in quasi-2D perovskites, and excitons will eventually undergo radiative recombination in high-<n> phases [13].

Table 1. A summary of performance of some advanced LED devices mentioned in the review.

Maximum Luminance	EL Peak (nm)	Peak EQE (%)	Stability	Ref.
		3D lead-based perovskite PeLEDs		
1680 cd m^{-2}	468	4.6	T_{50} at 1.8 mA cm^{-2} = 220 s	[14]
18,748 cd m^{-2}	487	13.6	not reported	[15]
8040 cd m^{-2}	513	21.3	not reported	[16]
103,286 cd m^{-2}	539	22.06	not reported	[17]
875 W sr^{-1} m^{-2}	787	19.6	T_{50} at 100 mA cm^{-2} = 20 h	[18]
663 W sr^{-1} m^{-2}	800	23.8	T_{50} at initial luminance of 107 W sr^{-1} m^{-2} = 32 h	[11]
		Quasi-2D lead-based perovskite PeLEDs		
1147 cd m^{-2}	483	18.65	T_{50} at 1 mA cm^{-2} = 1 min	[8]
2541 cd m^{-2}	486	16.7	T_{50} at initial luminance of 100 cd m^{-2} = 5 min	[19]
2150 cd m^{-2}	490	15.6	T_{50} at 0.44 mA cm^{-2} = 55.3 min	[20]
>5000 cd m^{-2}	495	25.6	T_{50} at 8 mA cm^{-2} = 115 min	[21]
73,896.6 cd m^{-2}	513	10.7	no reported	[22]
39,393 cd m^{-2}	524	30.84%	T_{50} at initial luminance of 100.88 cd m^{-2} = 100.6 min	[9]
1300 cd m^{-2}	680	25.8	T_{50} at initial luminance of 100 cd m^{-2} > 2000 s	[10]
		0D lead-based perovskite PeLEDs		
2916 cd m^{-2}	451	1.32	T_{50} at initial luminance of 100 cd m^{-2} = 1192 s	[23]
>10,000 cd m^{-2}	478	17.9	T_{50} at initial luminance of 100 cd m^{-2} = 2 h	[24]
∼5000 cd/m^{-2}	514	26.7	T_{50} at 1 mA cm^{-2} = 3 h	[25]
109,427 cd m^{-2}	518	23.45	T_{50} at 2.5 mA cm^{-2} = 77 s	[26]
470,000 cd m^{-2}	540	28.9	T_{50} at initial luminance of 1000 cd m^{-2} = 520 h	[27]
1500 cd m^{-2}	656	22.62	T_{50} at initial luminance of 100 ± 8 cd m^{-2} = 489 min	[28]
		Tin-based perovskite PeLEDs		
12 W sr^{-1} m^{-2}	894	8.3	T_{50} at 1 mA cm^{-2} ≈ 3 h	[29]
>1000 cd m^{-2}	630	20.29	T_{50} at initial luminance of 30 cd m^{-2} = 27.6 h	[30]
		OLEDs		
>10,000 cd m^{-2}	588	33.5	T_{50} at initial luminance of 1000 cd m^{-2} = 9.59 million h	[31]
		QLED		
≈3,300,000 cd m^{-2}	615	20.7	T_{50} at initial luminance of 1000 cd m^{-2} = 1.25 billion h	[32]

T_{50} means the time taken to reach 50% of the initial brightness

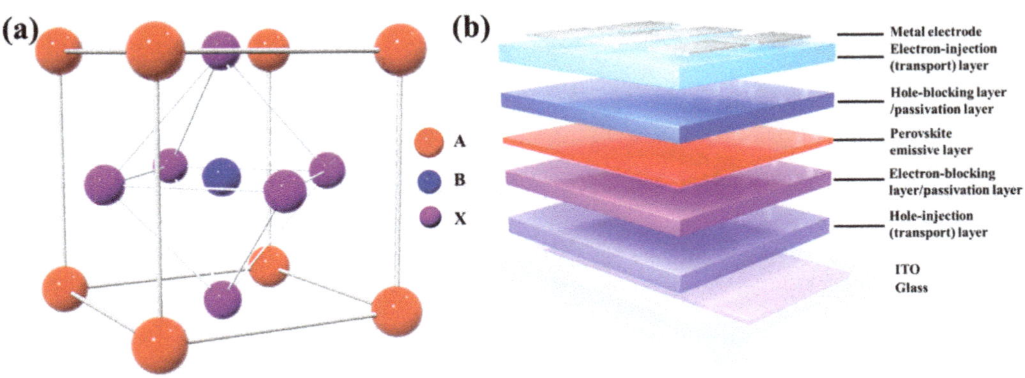

Figure 1. Illustrations of (**a**) crystal structure of perovskites, (**b**) device structure of a perovskite LED.

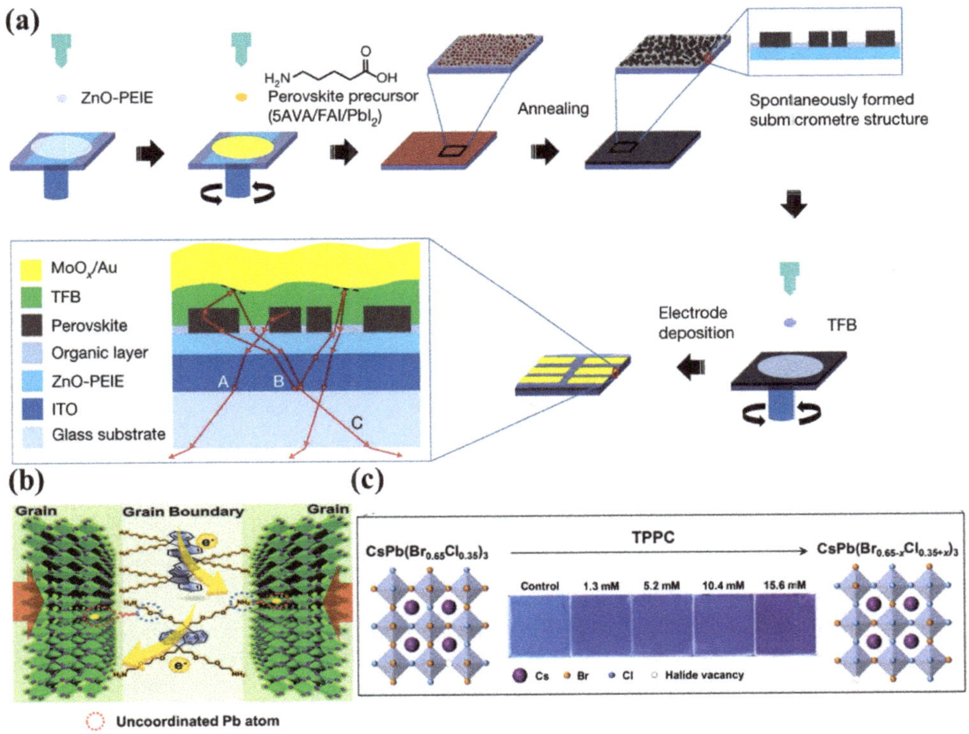

Figure 2. (**a**) The scheme of device fabrication with spontaneously formed submicrometer structure; arrows A, B and C represent light trapped in devices with a continuous emitting layer that can be extracted by the submicrometer structure or crystal structure of perovskites [33]. (**b**) Illustration of the defect passivation at grain boundary for perovskite film via Fl-OEGA [16]. (**c**) Mixed-halide perovskite film formed by an in situ halide ion exchange process [14].

One of the important performance parameters of PeLEDs is the EQE [12], which is expressed as follows: EQE = $\eta_{inj} \times \eta_{rad} \times \eta_{out}$. Here, η_{inj} refers to the efficiency of the charge injection from device electrodes into the perovskite emission layer. A bad energetic alignment and high trap state density at the interface will cause the nonradiative recombination of carriers and unbalanced injection of electrons and holes, leading to a lower η_{inj} [34–36]. η_{rad} is the radiative recombination efficiency of electron–hole pairs, which are directly related to the intrinsic property of emission layer materials, namely their PLQY. As for η_{out}, it is the outcoupling efficiency and reflects the ability of emitting photons generated from the perovskite emission layer into free space by the PeLED device. The value of η_{out} is affected by the degree of mismatch in refractive index between different device layers, where a large mismatch would yield severe optical losses via substrate and waveguide modes [37]. Hence, a perfect energetic alignment, a high-quality perovskite emission layer and interface and a well-matched refractive index between different device layers are essential to develop ideal PeLEDs with large EQEs [38].

In addition, the lifetime and stability of PeLEDs are also important for their commercial application. Compared with the nucleation and crystallization barriers of silicon semiconductors (470 kJ mol^{-1} and 280 kJ mol^{-1}, respectively), the crystallization of perovskites materials can be achieved at lower temperatures due to their smaller crystallization barriers (56.6–97.3 kJ mol^{-1}) [39]. However, because of the low formation energy and soft ionic property of perovskites, the quality of perovskite films fabricated by solution-processing are poor owing to many types of defects at the inter-grain boundaries, intra-grain boundaries and the surface of perovskites [40–43]. Under the influence of temperature, electric field, moisture, UV light and other external factors, these defects will lead to phase segregation of mixed-halide perovskites, phase transition, broadening and shifting of the peak emission, etc. [44–51]. Therefore, some reasonable strategies, including additive engineering, crystallization kinetic regulation, dimensionality control and interface modification, would be recommended to obtain a low defect density in the perovskite emission layer, which works in favor of the long-term and stable running of PeLEDs [36,52–56].

At the beginning of this review, basic information on perovskites materials and the device structures of PeLEDs as well as the factors influencing the EQE and stability of PeLEDs are discussed systematically. Then, the state of the art for lead-based PeLEDs is summarized, from bulk, quasi-2D to 0D perovskites materials being used as the emission layer due to lead-based perovskites' superior optoelectronic properties. Some novel PeLEDs based on co-friendly compositions, such as tin-based and double-metal perovskites, are also presented in Section 3. Subsequently, we discuss the challenges of PeLEDs that remain to overcome, including the low EQE of blue PeLEDs, poor device stability and EQE roll-off. In the end, a summary and outlook of PeLEDs in the future are provided.

2. Progress in Lead-Based PeLEDs

2.1. Progress in PeLEDs Based on 3D Lead-Based Perovskites

Three-dimensional $CsPbX_3$ perovskites are commonly utilized as emitters in PeLEDs owing to their easy synthesis, composition regulation and good charge-transport characteristics without the use of organic ligands [6,57–59]. Their films can be fabricated by spin coating, blade coating, slot-die coating and inkjet printing [60–64]. These solution-processing methods, however, generally produce inferior-quality perovskite films. Hence, some strategies, ranging from bulk passivation to surface/interface modification, have been developed to improve the film quality of perovskites [57,65,66]. Cao et al. [33] introduced 5-aminovaleric acid into perovskite precursor solutions. With the optimization of the feed ratio of 5-aminovaleric acid in the precursor solution, the additive not only passivated the surface defects but also increased the outcoupling efficiency via the submicrometer-scale structures formed in perovskite films (Figure 2a). Finally, the PeLEDs achieved an EQE exceeding 20% at 800 nm. Although organic molecules can passivate defects effectively via Lewis bases bearing amino, carboxyl, sulfonate, phosphate, thiol and other functional groups [67–73], the inferior conductibility of these organic additives weakens the transportation of carriers and reduces the EQE of PeLEDs [74,75]. A multifunctional organic semiconductor, 9,9′-bis(2-(2-(2-aminoethoxy)ethoxy)ethyl)fluorene (Fl-OEGA), was unitized as an additive by Song and coworkers [16]. The additive was mainly distributed at the grain boundaries, which released the residual stress and inhibited ion migration (Figure 2b). The corresponding green PeLED device obtained a top EQE of 21.3%. Mixing bromide/chloride is a facile and common strategy to realize blue light emission [76,77]; the low ion migration energy of Cl^{-1}, however, inevitably causes the generation of halide vacancies and phase separation [78,79]. In addition, the achievement of an emission wavelength < 470 nm based on 3D mixed Br/Cl perovskites is difficult due to the low solubility of chloride [80]. Wang et al. [14] designed an in situ anion exchange approach, where high-Cl-ratio and homogeneous $CsPbCl_xBr_{1-x}$ perovskites were obtained via treating $CsPbBr_3$ films with a solution of tetraphenylphosphonium chloride (Figure 2c). By additionally introducing the organic ammonium halide salts' passivator, deep-blue PeLEDs with the desired EQE were fabricated.

A benign interface quality is also essential for efficient PeLEDs, as it can improve band alignment and passivate surface defects and lead to rapid and balanced charge injection, restricted ion migration and nonradiative recombination, as well as ameliorated EQE roll-off and stability of the PeLEDs [81–84]. Wei et al. [17] dropped a phosphine oxide (PO-T2T) additive solution onto perovskites. The −P=O bond in the additive bonded tightly with the undercoordination Pb^{2+} via a dative bond at both the grain boundaries and the surface (Figure 3a). In addition, PO-T2T also formed a good energetic alignment to accelerate electron transportation between the perovskite layer and electron transport layer (Figure 3b). Finally, a peak EQE of 22.06% for green PeLEDs was obtained with a satisfactory maximum luminance (103,286 cd m^{-2}). Ion migration driven by the electric field is a notorious problem for PeLEDs, especially under high current density, resulting in EQE roll-off for 3D perovskite-based PeLEDs [79]. Zhao et al. [18] washed the top surface of perovskites with chloroform to eliminate residual PbI_2 and restrain ion migration. By further inserting an ultrathin poly(methyl methacrylate), a balance between carriers injection and increasing EQE was realized in the optimized PeLED device. The device could maintain >70% of the original EQE at 1400 mA cm^{-2} (Figure 3c). In recent years, the buried interface between the hole transport layer and the perovskites layer has drawn increasing attention, because it directly affects the growth of perovskites. Despite the existence of severe deep-level trap states at the buried interface, it is usually more difficult to modify than the top surface of perovskites. Carbazole-phosphonic acid derivatives are desirable hole transport layer materials with fast hole transfer rates and low interfacial trap state densities [85,86]. Di and coworkers [15] utilized [2-(9H-carbazol-9-yl)ethyl]phosphonic acid as the hole transport layer, in which the group of phosphonic acid could interact with Pb^{2+} to fill up halide vacancies and obtain a perfect buried interface. The corresponding blue PeLED devices had peak EQEs of 10.3 and 13.6% at 478 and 487 nm, respectively, with a fine maximum luminance (Figure 3d). Polymers are desirable modification materials for defect passivation and the suppression of perovskite decomposition induced by humidity [87–89]. Halpert and coworkers [90] reported a poly[(phenylglycidyl ether)-*co*-formaldehyde] (PCF)-protected perovskite layer as the emitter for PeLEDs (Figure 3e). The ether in PCF was able to form hydrogen bonds with methylammonium cations and replace halide vacancy via coordination bonds to increase the perovskite film quality. Meanwhile, PCF also reduced the invasion of moisture under a >70% RH environment, which was beneficial for the practical application of PeLEDs.

2.2. Progress in PeLEDs Based on Quasi-2D Lead-Based Perovskites

Since the pioneering work reported by Calabrese et al. [91], quasi-2D perovskites have been one of the most promising materials for photoelectronic applications resulting from their unique structural characteristics [92]. They are formed through inserting bulky organic cations into the A site of the 3D perovskite lattice. The general chemical formula of quasi-2D perovskites is $A'_2A_{n-1}B_nX_{3n+1}$, where the element composition of A, B and X is similar to that of 3D perovskites. In addition, A' refers to the organic spacer cation and n indicates the number of $[BX6]^{4-}$ octahedral slab sandwiched between organic spacer cations [93]. Because of the small value of dielectric constants for organic spacer cations, a "quantum-well" structure can be produced for quasi-2D perovskites [94,95]. The quantum and dielectric confinement effects originating from the "quantum-well" structure contribute to the large exciton binding energy (E_b), in combination with the tunable bandgap by changing the thickness of the "quantum-well", and high PLQY due to the efficient energy transfer from low n-phase to high n-phase; quasi-2D perovskites are thus very suitable emission layer materials for PeLEDs [96].

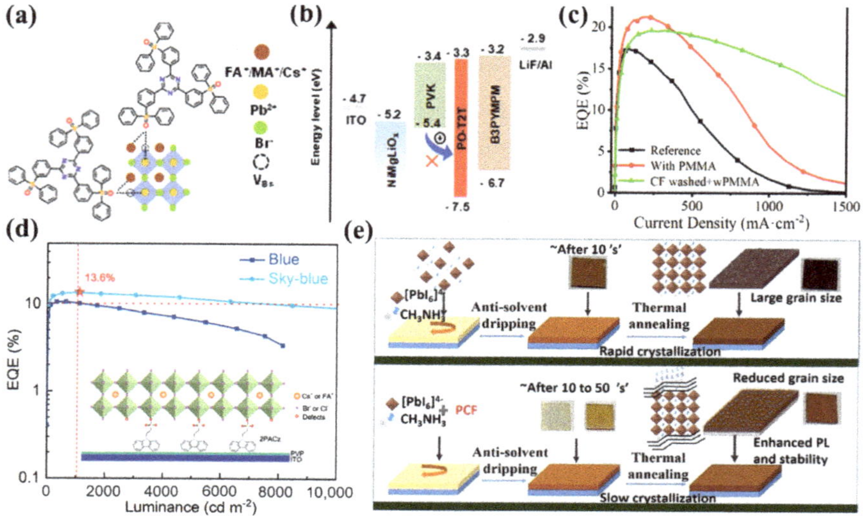

Figure 3. (**a**) The interaction between PO-T2T and perovskites [17]. (**b**) The energy level alignment in PeLEDs [17]. (**c**) EQE–current density relations of the reference, with PMMA layer and CF washed + PMMA device [18]. (**d**) The evolution curves of EQE with the change in luminance (inset: the modification of carbazole-phosphonic acid at buried interface) [15]. (**e**) Illustration of $MAPbI_3$ formation through one-step method without (**top**) and with (**below**) protection of PCF [90].

An acceptable quality of quasi-2D perovskite films is the basic requirement for ideal PeLEDs. Tang et al. [8] added the bifunctional benzoic acid potassium into the percussor solution of mixed-halide quasi-2D perovskites, in which the benzoic acid (BA^-) group could serve as a Lewis base to interact with undercoordinated Pb^{2+}, and K^+ could bind tightly with halide atoms via ionic bonds, leading to the thorough modification of quasi-2D perovskites with a reduction in defect-induced nonradiative recombination and halide ion migration (Figure 4a). A record EQE for sky-blue PeLEDs was achieved, and the work guided the selection of appropriate additives employed in perovskite materials. Though the PLQY of quasi-2D perovskites is high resulting from their special properties, as previously mentioned, the undesirable electrical conductivity of organic spacer cations will retard the charge injection and undermine the EQE of PeLEDs [97]. Sirringhaus and co-workers [98] displaced organic spacer cations with sodium cations. The fabricated perovskite layer had a fine orientation. In addition, an amorphous $NaPbBr_3$ was formed with the incorporation of sodium salt, resulting in the generation of quasi-2D perovskites. By optimizing the feed ratio of sodium salt and a little organic additive, the green PeLEDs obtained an EQE of 15.9% and a good operational lifetime (Figure 4b).

Because the formation energy of different n phases of quasi-2D perovskites is similar, there is a wide n phase distribution in quasi-2D perovskite materials [99]. Although the cascade energy transfer from high-n phase to low-n phase enables the improvement of the PLQY, a too wide n phase distribution is unfavorable due to the low color purity and possible energy loss in energy transfer. In addition, a low-n (n = 1) phase causes strong electron–phonon coupling interactions, which cause nonradiative recombination [100–102]. Hence, some strategies have been developed to narrow the n phase distribution of quasi-2D perovskites. Sargent and coworkers [21] synthesized a bifunctional additive tris(4-

fluorophenyl)phosphine oxide (TFPPO) and dissolved it into antisolvent to treat perovskites during the fabrication of LED devices. The strongly electronegative fluorine atoms in the additive interacted with the organic ammonium cations through hydrogen bonds to restrict the low-n phase formation and achieve a monodispersed "quantum-well" structure. In addition, the phosphate could fill up the unsaturated sites to reduce the trap-state density (Figure 4c). Finally, the authors realized the desired EQE value of 25.6% for green PeLEDs while exceeding the 2 h half-life. The interaction intensity between the organic spacer cation and Pb^{2+} directly influence the crystallization kinetics and n-phase distribution of quasi-2D perovskites. Yan et al. [20] investigated the effect of Pb^{2+}'s ability to coordinate a series of zwitterion spacers on different n-phase formations (Figure 4d). The results showed that the stronger coordination ability the Pb^{2+} zwitterions have, the easier a low-n phase will be obtained. Therefore, zwitterions with moderate coordination ability are able produce a concentrated n-phase distribution of quasi-2D perovskites, and were used to fabricate a high-performance blue PeLED with a 15.9% EQE. Apart from controlling the n-phase distribution, the inhibition of phase separation induced by halide migration in mixed-halide quasi-2D perovskites is also necessary. Yang et al. [103] comprehensively considered the heteroatom, conjugation length, and molecular twist of organic cation spacers, and they found that twisted and extended conjugated organic cation spacers enabled a narrow n-phase distribution and a negligible segregation of mixed halides (Figure 4e). Finally, they reported a number of stable and high-efficiency PeLED devices with emission wavelengths ranging from the red to NIR regions.

Interfacial engineering between quasi-2D perovskites and charge injection layers is an efficient strategy to ameliorate the performance of PeLEDs [53]. You's group [19] utilized CsCl to modify the hole injection layer of poly(3,4-ethylenedioxythiophene) polystyrene sulfonate (PEDOT:PSS). The incorporation of CsCl could achieve an oriented alignment of PEDOT chains and change the position of the valance band. These modifications were beneficial to the improvement of carrier injection and transportation (Figure 5a). Moreover, according to the results of the XRD and transient absorption spectra, the formation of the low-n phase was also restrained by the modified hole injection layer (Figure 5b). The optimized sky-blue quasi-2D perovskite PeLEDs obtained an 16.07% EQE and a satisfactory lifetime. Chueh et al. [22] added a conductive interlayer between the electron transport layer and perovskite emission layer, where undesirable energy transfer and nonradiative recombination were inhibited (Figure 5c). They found that the modification can significantly improve the brightness of PeLEDs and alleviate EQE roll-off at high current densities. The refractive index for most organic hole injection layer materials is very small compared with that of perovskite materials, and this large difference will cause the photons to be captured by the hole injection layers [104,105]. To overcome this photon loss, Xie et al. [9] utilized NiO_x as the hole injection layer materials, which possesses the merits of adjustable energy levels and refractive index [106,107]. By using a suitable doping ratio of Mg in NiO_x film and inserting a polymer layer before the deposition of the quasi-2D perovskite layer, the optimized green PeLEDs' outcoupling efficiency (41.82%) was effectively improved as well as (Figure 5d) obtaining a rapid and balanced charge carrier injection. Finally, a record EQE of 30.84% was achieved for the green PeLEDs (Figure 5e).

2.3. *Progress in PeLEDs Based on 0D Lead-Based Perovskites*

Zero-dimensional perovskites, namely perovskite quantum dots (PsQDs), can be understood as a core of stoichiometric ABX_3 covered by a BX_2 inner shell and an $A'X'$ outer shell, where A' refers to either Cs^+ or the ligand cation (normally alkylammonium ions), and X' consists of halide anions and ligand anions (normally alkyl carboxylate ion) [108]. With the confinement of the surface ligand, the PeQD possesses some unique virtues, including adjustment of the emission wavelength according to size and composition control, high color purity and a near-unity PLQY [109–111]. Since the first report of the synthesis of a PeQD colloidal solution using the ligand-assisted reprecipitation method (LARP) in 2014 [112], some significant progress has been made in PeQD-based LEDs (PeQD LEDs), such as exceeding an EQE of 20% for red and green LEDs, as well as exceeding 10% EQE for blue LEDs [59,113,114].

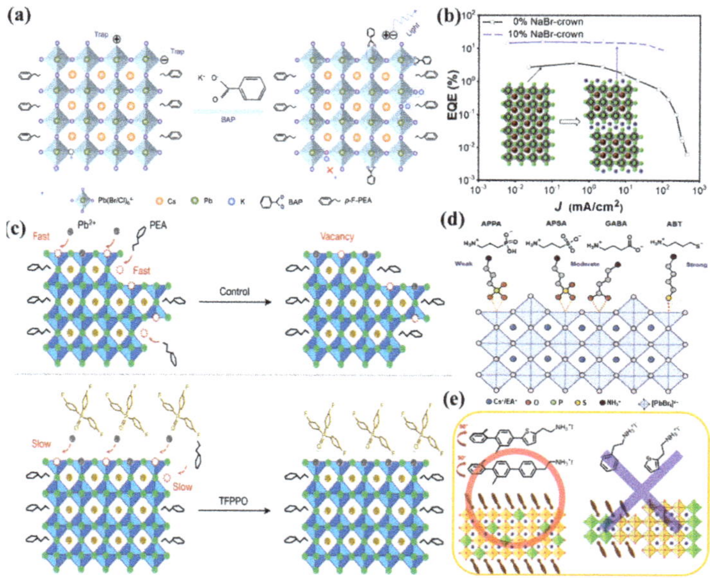

Figure 4. (**a**) The passivation function of additive in quasi-2D perovskites [8]. (**b**) EQE–*J* curves of PeLEDs with and without addition of NaBr [98]. (**c**) The process of crystallization for quasi-2D perovskites after the antisolvent treatment with and without TFPPO [21]. (**d**) The interaction intensity between different zwitterions and perovskites [20]. (**e**) Phase distribution in mixed-halide quasi-2D perovskites with different organic cation spacers [103].

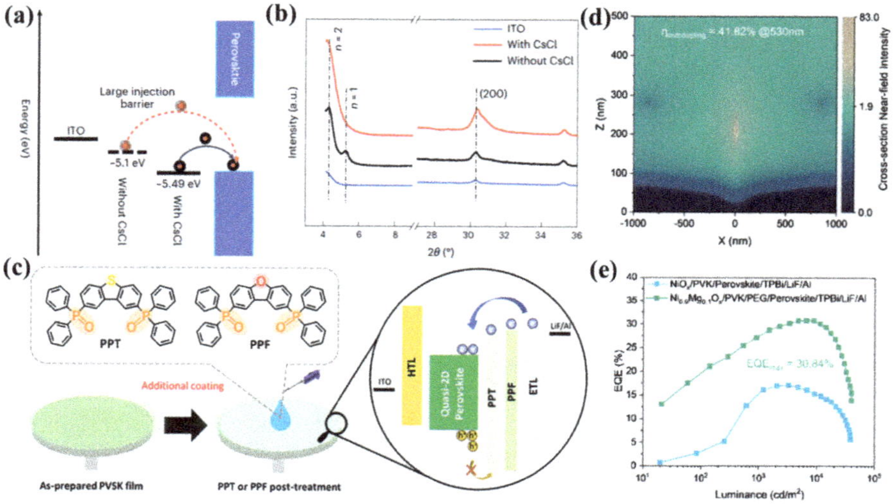

Figure 5. (**a**) Band alignment in PeLEDs using PEDOT:PSS with and without CsCl addition [19]. (**b**) XRD results of perovskite films deposited on PEDOT:PSS substrates with and without CsCl addition [19]. (**c**) Band alignment in PeLEDs with interfacial post-treatment [22]. (**d**) Cross-section near-field intensity distribution of the optimized PeLED device for light outcoupling at 530 nm [9]. (**e**) EQE–luminance curves of reference and modified PeLEDs [9].

Despite the promising potential of PeQDs in LED applications, the highly dynamic adsorption and desorption of surface ligands for PeQDs will influence their stability and PLQY during the process of purification, film formation and long-term storage [115]. In addition, ligands with long alkyl chains, such as the commonly used oleic acid (OA) and oleylamine (OAm), have lower conductibility and poor interaction between PeQDs [116]. These shortages limit the practical applications of PeQD LEDs. Hence, the synthesis of bright and stable PeQD colloidal solutions is the primary requirement for the corresponding LEDs and other optoelectronic device applications. Considering the ease of ligand desorption due to the proton transfer between OA and OAm, Sargent et al. [117] designed a novel amine-free synthetic approach, in which CsOA and PbOA reacted with tetraoctylammonium halides, and obtained a PeQD colloidal solution exhibiting satisfactory stability during the purification process (Figure 6a). Based on the PeQD, the fabricated green and blue LEDs showed several-fold improvement in EQE compared with traditional OA/OAm-capped PeQD LEDs. Kovalenko and coworkers [118] also changed the ligand type used in the hot injection synthesis method of PeQDs. They added didodecyldi-methylammonium halides, frequently applied ligands in the post-treatment process of PeQD colloidal solutions, into the precursor solution as the sole ligand to obtain $CsPb(Br_{1-x}Cl_x)_3$ PeQDs (Figure 6b). Benefiting from shorter alkyl chains compared with OA/OAm and an optimized device structure, efficient pure blue LEDs were achieved. Zeng et al. [119] found that dodecylbenzene sulfonic acid could act as a "Br-equivalent" ligand to bind tightly with Pb^{2+} due to the strong ionic nature of the sulfonate group. This inhibited not only the desorption of ligands but also the formation of halide vacancy (Figure 6c) Using the perfect ligand, the synthesized PeQD colloidal solution could easily achieve a PLQY > 90% and maintain the desired stability even after eight purification cycles or exceeding 5 months of storage, which has promising applications in LEDs.

Ligand exchange is also an effective strategy for modifying the stability and photoelectric properties of PeQD colloidal solutions [59,120,121]. Bakr and coworkers [122] added OA and di-dodecyl dimethyl ammonium bromide into a purified PeQD solution. They proposed that the OA could induce the desorption of protonated OAm, and then the PeQD surface would be reconstruction by the further substitution of OA with di-dodecyl dimethyl ammonium bromide. With the success of ligand exchange, they were able to achieve a green LED based on a halide–ion-pair-capped PeQD (Figure 7a). Zhang et al. [26] found that dimethyl butylamine and ethyl bromide reacted spontaneously at room temperature via nucleophilic substitution to yield alkylammonium bromide. Therefore, they added the two additives into a purified PeQD solution. The alkylammonium and Br^- filled up the defects at the A and X sites, respectively, and partially replaced the OA/OAm ligand, leading to an improvement in PLQY, stability and electroconductivity of the PeQDs (Figure 7b). This in situ ligand compensation approach enabled the corresponding green PeQD LEDs to acquire an EQE of 23.45% and 109,427 cd m^{-2} of maximum luminance. As mentioned above, PeQDs prefer to bind with strong polar ligands due to the soft ion nature of perovskites [123–126]. These ligands cannot dissolve in non-polar solvents; high polar solvents, however, undermine the stability of PeQDs [59,127–129]. Based on this situation, a liquid Iodotrimethylsilane ligand was used to treat the $CsPbI_3$ colloidal solution [28]. The ligand possessed a fine solubility in non-polar solvents, and it was able to react with oleate to etch PeQD surface in situ via the produced HI, resulting in the reduction in defect density and improvement in stability due to the strongly binding ligands. A 23% EQE for red PeQD LEDs was finally reported (Figure 7c), and this ligand has a similar effect for other compositional PeQDs. Compared with the organic cap layer, the inorganic shells have more advantages, for example, good conductivity and stability [121]. Lee et al. [23] utilized $CsPbBr_3$ PeQDs as seeds, and through blending the seeds with $ZnCl_2$, $ZnBr_2$ and S-ODE, $CsPb(Br_{1-x}Cl_x)_3$ PeQDs covered with a ZnS shell were obtained due to low lattice mismatch (Figure 7d). The ZnS shell protected the $CsPb(Br_{1-x}Cl_x)_3$ core from the external effect, accelerated the carrier injection and restricted the ion migration of mixed halides. Finally, they achieved a 1.32% EQE for deep-blue LEDs with a satisfying operational lifetime.

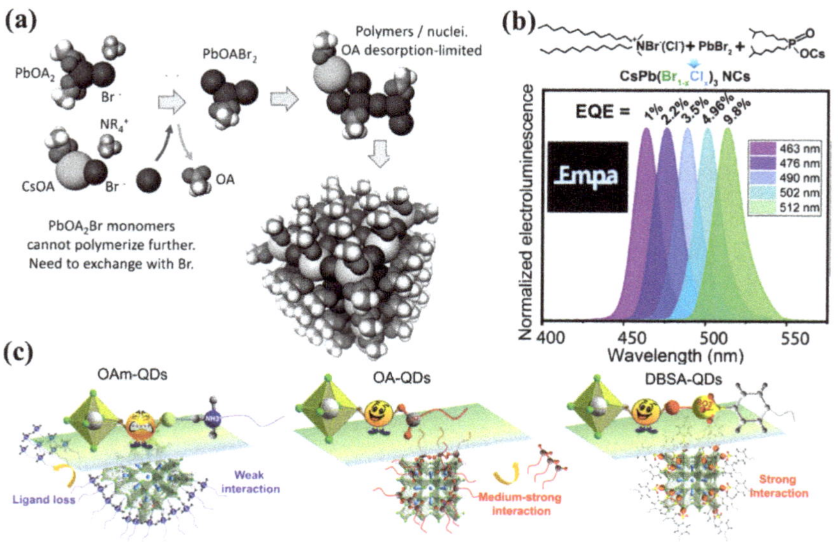

Figure 6. (**a**) Molecular dynamics simulation of $CsPbBr_3$ nucleation and growth steps without amine ligand using density functional theory [117]. (**b**) Normalized PL spectra of $CsPb(Br_{1-x}Cl_x)_3$ PeQDs with didodecyldi-methylammonium ligand [118]. (**c**) The interaction intensity between PeQDs with ligands with different binding motifs ($-NH_3^+$ (**Left**), $-COO^-$ (**mid**), $-SO_4^+$ (**Right**)) [119].

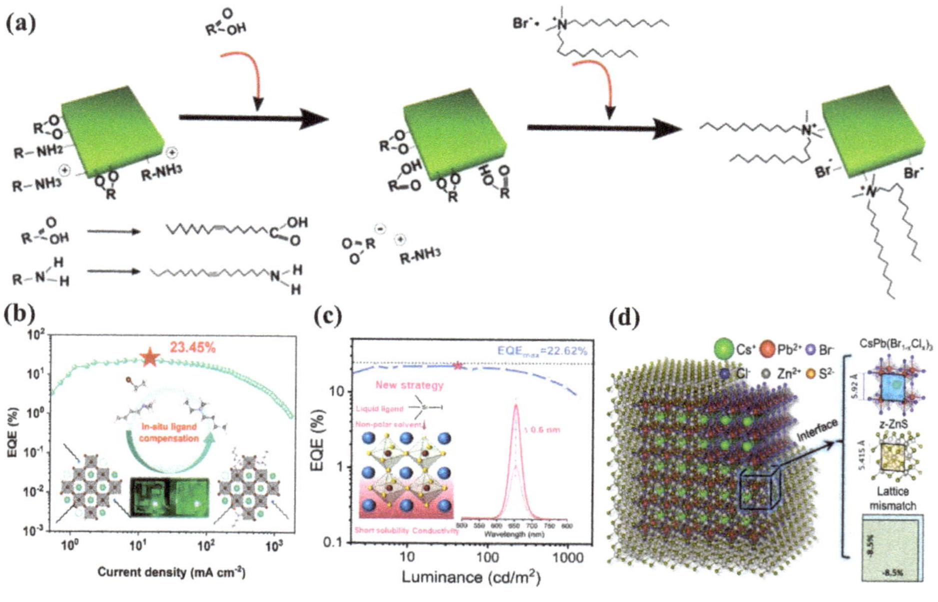

Figure 7. (**a**) The mechanism of ligand exchange on surface of PeQD [122]. (**b**) Illustration of in situ ligand compensation for PQDs and the EQE evolution with current density of optimized PeLEDs [26]. (**c**) EQE–luminance curves of PeLED device based on Iodotrimethylsilane-ligand-treated PeQD [28]. (**d**) Simulated 3D atomic model of $CsPbBr_3$ PeQDs covered by ZnS [23].

Although PeQD films have lower defect state densities than polycrystalline bulk films with the existence of surface ligands, there are also several undercoordinated Pb^{2+} molecules at the surface of PeQD films due to the easy desorption and steric effects of ligands, so a post-treatment of PeQD films is necessary to obtain a high-EQE device [130,131]. Lee et al. [113] first doped guanidinium into $FAPbBr_3$ to increase the stability and charge confinement effect of PeQDs. On the other hand, passivation interlayers (TBTB) were deposited on the top surface of PeQD films to further fix the bromide vacancy defects, which ameliorated the photoluminescence properties of PeQD films and the carrier injection balance of PeLEDs (Figure 8a). A 23.4% EQE and 108 cd A^{-1} current efficiency were obtained in the optimized PeLED device (Figure 8b). Brovelli et al. [25] investigated the influence of emission layer thickness in the balancing charge transport by further inserting the NiO_x layer in both the top surface and in the buried interface to optimize the energetic alignment (Figure 8c). They fabricated a stable green PeQD LED with an EQE as high as 26.7%. The selection of suitable charge transportation layer materials is crucial for the formation of benign interfaces without inserting extra and complex modification materials. TPBi is a commonly used electron transportation layer material; however, it has a lower electron mobility than the hole mobility of materials generally used as hole transportation layer materials (such as PTAA and poly-TPD) [132–135]. Considering this case, Xu et al. [136] synthesized new electron transportation layer materials, denoted as B2, to facilitate electron injection. The materials were able to perfectly balance the carrier injections and suppress nonradiative recombination at the top surface (Figure 8d), and a 13.17% EQE for blue PeQD LEDs was achieved with an 8656.67 cd·m^{-2} maximum luminance.

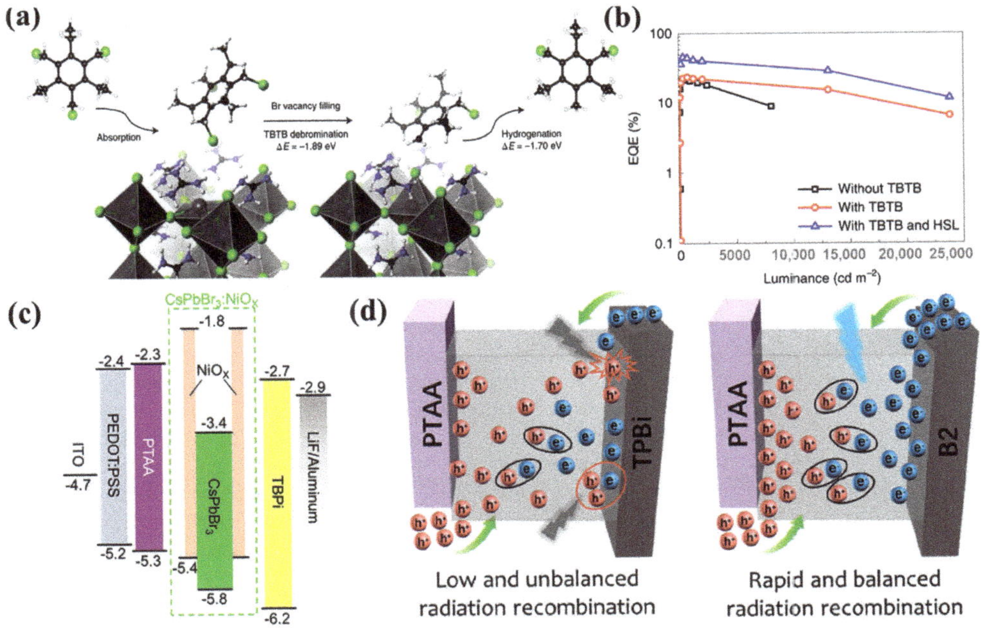

Figure 8. (**a**) The bromide vacancy healing process using TBTB molecules on guanidinium-terminated $FAPbBr_3$ surface simulated using density functional theory [113]. (**b**) EQE–luminance curves of PeLEDs with or without a TBTB interlayer and a hemispherical lens (HSL) [113]. (**c**) Energy band alignment of PeLEDs with NiO_x treatment [25]. (**d**) Illustration of the charge recombination in PeLED devices based on TPBi (**left**) and B2 (**right**) [136].

Except for the commonly used hot injection and ligand-assisted reprecipitation methods, in situ synthesis of PeQDs in the substrate is a simpler and more favorable method to decrease the cost of LED device fabrication [27,137,138]. Yuan et al. [24] manipulated the chemical structure of phenylethyl ammonium ligands and successfully produced monodispersed PeQDs on the hole transportation layers in situ. On the one hand, extra methyl substitution in the head group inhibited the yield of the layered perovskites; on the other hand, halide substitution at the ortho-site of the phenyl ring effectively increased the interaction between the ligand and perovskites (Figure 9a). Therefore, the size of the PeQDs could be adjusted continuously through changing the ligand concentration, and the short and conjugated ligands enabled the suitable coupling between PeQDs (Figure 9b). This strategy resulted in high-performance LEDs with diverse emission wavelengths based on the size and composition of the PeQDs (Figure 9c).

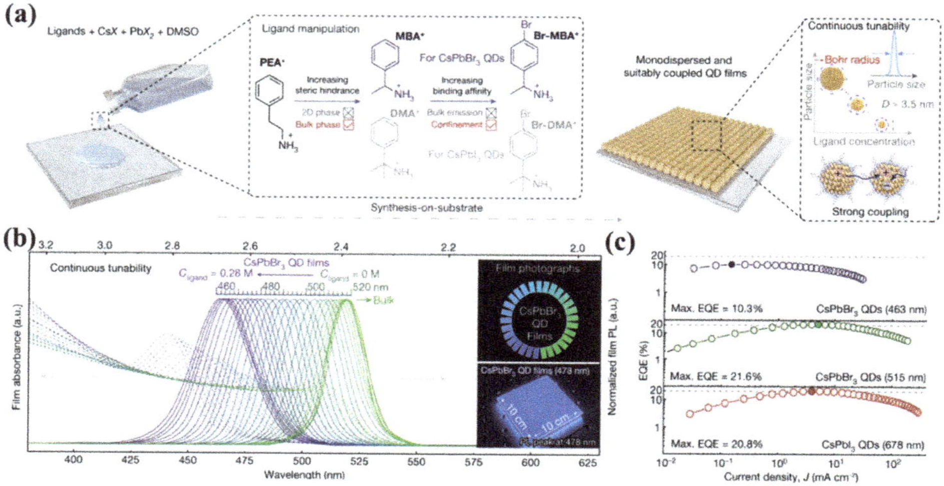

Figure 9. (**a**) Ligand design for in situ synthesis of monodispersed and suitably coupled PeQDs on substrate. (**b**) Photoluminescence and absorption spectra of $CsPbBr_3$ QD films with different ligand concentrations. (**c**) The evolution of EQE with current density of diverse colors of PeLEDs [24].

3. Progress in Lead-Free PeLEDs

Although EQEs near the milestone of 30% for lead-based PeLEDs has been reported, the toxicity of lead for humans and ecosystems has been a concern for a long time. Researchers have contributed tremendous efforts to find reasonable substitute elements for lead without decreasing the EQE and stability of PeLEDs [139]. Having the same main group element, the ionic radius and valence electron configuration of tin are similar to lead, and hence tin-based perovskites are viewed as very ideal and eco-friendly substitute materials for the emission layer of PeLEDs [140,141]. Friend et al. [142] demonstrated the application of $CH_3NH_3SnI_3$ for PeLEDs emitting at NIR in 2016 due to the smaller bandgap of $CH_3NH_3SnI_3$ compared to $CH_3NH_3PbI_3$. Meanwhile, with the increase in the Br ratio, the emission wavelength of PeLEDs could blue shift down to 667 nm, which widens the application of PeLEDs in sensing, medical devices and optical communication (Figure 10a). Despite the merits of tin-based perovskites mentioned above, the performance of their LEDs nowadays is far behind that of lead-based perovskite LEDs. The major reasons for this are their rapid crystallization kinetics and the easy oxidization of Sn^{2+} into Sn^{4+}, leading to inferior film quality [143,144]. Huang et al. [29] investigated the crystallization process of tin-based perovskites via in situ PL spectra, and they found that the fast aggregation of clusters at the primary growth procedure was effectively retarded through the strong chemical interactions between SnI_2 and additives (PEAI and vitamin B1). A good film quality with improved crystallinity and luminescence efficiency was obtained (Figure 10b). Finally, the best EQE of 8.3% was achieved in tin-based PeLEDs emitting at 894 nm (Figure 10c). Wang and coworkers [30] introduced cyanuric acid (CA) into the precursor solution of quasi-2D tin-based perovskites. The C=N in CA could strongly coordinate with Sn^{2+} to restrict the oxidization of Sn^{2+}, and the hydrogen bond interaction between −OH and I^- reduced the loss of I^-. In addition, the dimers or trimers for the mixed keto and enol tautomers of CA were orderly arranged at the surface of the perovskites, facilitating the perpendicular growth of perovskites (Figure 10d,e). These advantages enabled a 20.29% EQE of tin-based PeLEDs to be obtained, which is comparable to high-performance lead-based PeLEDs (Figure 10f).

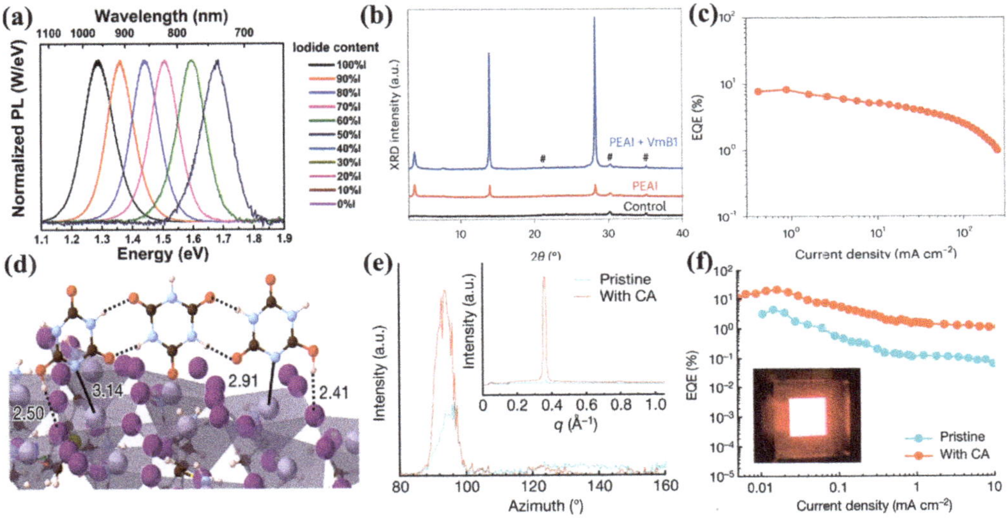

Figure 10. (**a**) The evolution of PL spectra of perovskite with change of iodide content [142]. (**b**) XRD patterns of perovskite film with different additives; ♯ indicates the diffraction data from the ITO substrate [29]. (**c**) The curve of EQE with the change in current density for PeLEDs [29]. (**d**) The simulated stable configurations for the tautomeric CA trimer on the surface of tin-based perovskites [30]. (**e**) Intensity curves of GIWAXS patterns along the (002) ring for the samples with and without CA (inset: integrated intensity of GIWAXS patterns for the samples) [30]. (**f**) The curve of EQE with the change in current density for PeLEDs with and without CA [30].

The other alternative of lead-based perovskites is double perovskites, whose universal chemical formula of it is $A_2B^IB^{III}X_6$, where Pb^{2+} is replaced by a combination of monovalent metal cations (Na^+, K^+, Ag^+, etc.) and trivalent metal cations (In^{3+}, Sb^{3+}, Bi^{3+}, etc.) [145,146]. The most widely studied double perovskites, like $Cs_2(Ag/Na/K)(Bi/In)X_6$, not only have lower toxicity, but also higher stability and carrier lifetimes than those of lead-based perovskites [139,147]. In addition, these double perovskites can produce broad emission wavelengths originating from the interaction between the carrier and the soft octahedron lattice, showing prospective applications for white LEDs based on a single emitter [148–150]. Tang et al. [151] employed theoretical calculations to study the luminescent property of $Cs_2AgInCl_6$, and they found the parity-forbidden cause of the extremely small value of PLQY. To break the forbidden parity, Na^+ was incorporated into $Cs_2AgInCl_6$, which achieved a remarkable improvement in PLQY. By further slight doping with bismuth, the fabricated electroluminescence device obtained 86 ± 5% of quantum efficiency with a long operational lifetime (exceeding 1000 h) (Figure 11a,b). Qu et al. [152] performed density functional theory calculations and found that Bi-doping also enabled the breaking of the parity-forbidden transition of $Cs_2AgInCl_6$ to ameliorate the PLQY. Therefore, $Cs_2AgIn_{0.9}Bi_{0.1}Cl_6$ quantum dots were synthesized and used as the emitter materials for

LEDs. The optimized white LEDs had a 0.08% EQE with competitive brightness and long-term stability (Figure 11c).

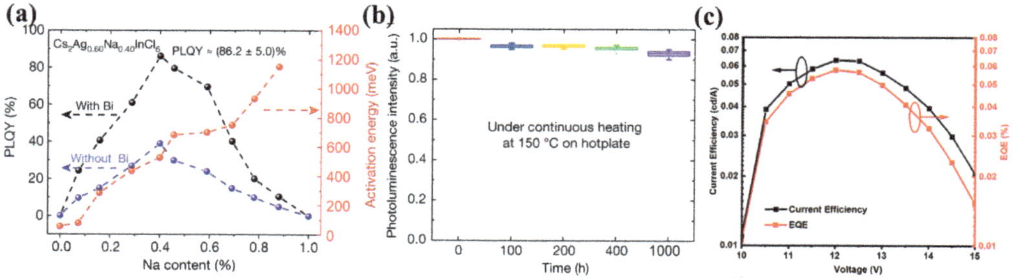

Figure 11. (**a**) Dependence of activation energy and PLQY of $Cs_2Ag_xNa_{1-x}InCl_6$ with and without doping of Bi on Na content [151]. (**b**) The evolution of PL intensity of $Cs_2Ag_{0.60}Na_{0.40}InCl_6$ under continuous heating at 150 °C on a hotplate, measured after cooling to room temperature [151]. (**c**) The evolution of EQE and current efficiency according to the increase in driving voltage [152].

4. Challenges for PeLEDs

4.1. Poor Devices' Operational Stability

Perfect operational stability is a basic requirement for the practical application of PeLEDs, compared with the lifetime of organic LEDs (T_{50} > 9,500,000 h at 1000 cd m^{-2}) and quantum-dot LEDs (T_{50} = 125,000,000 h at 100 cd m^{-2}) [31,32]. the record T_{50} for PeLEDs at an original brightness of 100 cd m^{-2} is only 2500 h [153]. The instability mainly originates from the perovskite layer as well as its top and buried interfaces [154]. Bulk perovskite materials are sensitive to certain external factors. The temperature of PeLEDs will constantly rise with the consecutive running of these devices, and the increased temperature may cause the dissociation of excitons into free carriers. Even worse, a too high temperature can decompose the perovskites, especially for organic and inorganic hybrid perovskites with organic cation at the A site [155–157]. The water and oxygen in the ambient may also influence the stability of PeLEDs [158]. The water can migrate inside the perovskite lattice via surface defect sites to form hydrates, and these hydrates enable the formation of deep-level defects and the decomposition of perovskites [51,159,160]. The oxygen can trap hole carriers to produce reactive superoxide, which may cause the oxidization of halide anions and the degradation of organic A site cations [47]. The electric field has a significant influence on the stability of PeLEDs through halide segregation, band bending, electrode corrosion, etc. [161–163]. In addition, a poor interface quality will also hurt the stable operation of PeLEDs; for example, high trap-state density at the interface leads to the restriction of charge transportation, and the accumulative charge at the interface may cause the degradation of the perovskite layer [164].

4.2. EQE Roll-Off

EQE roll-off is a general problem for all types of LED devices, especially in PeLEDs. It refers to the rapid decrease in EQE when LEDs are operated at a high current density. The major factors of EQE roll-off are unbalanced charge injection, Joule heating and Auger recombination [165–168]. Unbalanced charge injection can be ascribed to the inferior interface quality and inappropriate band alignment, and the unbalanced charge injection causes carrier accumulation, leading to the aggravation of Auger recombination and Joule heating [17,154]. In addition, Auger recombination probability is proportional to exciton binding energy, so there is severe Auger recombination in strong-confinement PeQD and quasi-2D perovskites [169,170].

4.3. Low EQE for Blue PeLEDs

Although nowadays red and green PeLEDs have achieved EQEs exceeding 20%, the EQE for their blue counterparts (emission wavelength < 470 nm) is very low (~12%). Mixed Cl-Br perovskites are the most straightforward materials for blue emission; however, ion migration and phase separation notoriously take place in these materials, especially under high bias voltages, leading to the reduction in the EQE and a shift in emission wavelength [137,171,172]. Another candidate is low-n phase (n = 1, 2) quasi-2D perovskites due to their special "quantum well" structure; however, there are strong electron–phonon coupling interactions in low-n phase quasi-2D perovskites, which undermine the luminescence efficiency of PeLEDs [173]. In addition, $CsPbBr_3$ quantum dots of an ultra-small size can also realize blue light emission according to the quantum confinement effect; nevertheless, the synthesis of the ultra-small-size quantum dots is difficult to achieve without comprising their stability and efficiency, because the smaller the quantum dots are, the larger the surface-to-volume ratio is, so higher ligand densities are needed, resulting in an inefficient carrier injection and undesirable EQE for PeLEDs [59,174,175].

5. Summary and Outlook

This review summarizes the recent progress in the application of lead-based perovskite materials in LEDs. Various strategies have been developed to improve the performance and long-term storage and operational stability of PeLED devices, including additive engineering, interfacial engineering, dimensionality control, device structure optimization, etc., which allow PeLEDs to show a comparable EQE with advanced organic and quantum-dot LEDs. Here, we provide some perspectives on the future development of this interesting field:

1. Considering commercial applications, further breakthroughs in the performance of PeLEDs are needed, especially in their operational stability. Polymer materials may act as good additives for the enhancement of the optoelectronic properties and stability of the perovskite emission layer [176,177]. On the one hand, these polymers can fill the vacancies at the grain boundaries of perovskites due to their rich functional groups, and the interaction between these functional groups and halide anions or Pb^{2-} enables the modulation of the orientation and crystallization kinetics of perovskites [178–180]. In addition, the polymer network can also protect the perovskites against ion migration and degradation induced by moisture, temperature and other external stimuli [90]. The extra insertion of the passivator layer is a prevalent method to modify the charge injection and interface quality; however, it increases the complexity of device fabrication. Therefore, the reasonable design of novel charge transportation materials is desirable to achieve efficient carrier injection and reduce trap-state density at the surface of perovskites; in addition, a suitable selection of charge transportation materials benefits the improvement of heat dissipation and outcoupling efficiency [15,136].

2. Despite the significant success in red and green PeLEDs, there is still a challenge in obtaining their blue counterparts. To broaden the applications of PeLEDs in displays and improved luminance, further optimization of blue PeLEDs with a satisfying performance is necessary. Firstly, suitable additives are key to inhibiting the formation of Cl vacancy defects and phase separation in mixed Br/Cl perovskite emitters. Alkali metal ions are potential candidates to fix halide anions via strong ion bonds [55,181], and conjugated benzene rings with sulfonate and phosphate groups can bind with undercoordinated Pb^{2+} effectively and modulate the crystallization process of perovskites [24,182]. Hence, choosing additives based on these considerations is recommended. As for pure bromide-based PeQD emitters, some surface ligands with the characteristics mentioned above are also beneficial for efficient charge injection. In addition, there is a requirement to develop novel hole transportation materials due to inefficient hole injection originating from unmatched band alignment [183–185].

3. Lead is toxic to humans, animals and the environment; hence, finding safe and high-performance lead-free PeLEDs is also important. Nowadays, tin-based PeLEDs are the

most likely alternatives for their lead-based counterparts; however, the rapid crystallization and easy oxidation of tin-based perovskites tend to cause the formation of deep-level defects (like Sn^{4+}). Organic molecules possessing appropriate coordination ability with Sn^{2+} are always used to retard the crystallization kinetics and suppress oxidation of Sn^{2+} in tin-based perovskites, and combined with a small addition of reductive reagent, fine tin-based perovskite films will be obtained to achieve efficient PeLEDs [29,30].

Funding: This work was financially supported by the National Natural Science Foundation of China (No. 52373246) and the Shanxi Basic Research Program Project (No. 20210302123128).

Institutional Review Board Statement: Not applicable.

Informed Consent Statement: Not applicable.

Data Availability Statement: The data can be available upon request to the corresponding author.

Conflicts of Interest: The authors declare no conflict of interest.

References

1. Alcocer, M.J.P.; Leijtens, T.; Herz, L.M.; Petrozza, A.; Snaith, H.J. Electron-Hole Diffusion Lengths Exceeding Trihalide Perovskite Absorber. *Science* **2013**, *342*, 341–344. [CrossRef]
2. Sutherland, B.R.; Sargent, E.H. Perovskite Photonic Sources. *Nat. Photonics* **2016**, *10*, 295–302. [CrossRef]
3. Quan, L.N.; Rand, B.P.; Friend, R.H.; Mhaisalkar, S.G.; Lee, T.W.; Sargent, E.H. Perovskites for Next-Generation Optical Sources. *Chem. Rev.* **2019**, *119*, 7444–7477. [CrossRef] [PubMed]
4. Protesescu, L.; Yakunin, S.; Bodnarchuk, M.I.; Krieg, F.; Caputo, R.; Hendon, C.H.; Yang, R.X.; Walsh, A.; Kovalenko, M.V. Nanocrystals of Cesium Lead Halide Perovskites (CsPbX3, X = Cl, Br, and I): Novel Optoelectronic Materials Showing Bright Emission with Wide Color Gamut. *Nano Lett.* **2015**, *15*, 3692–3696. [CrossRef] [PubMed]
5. Lu, M.; Zhang, Y.; Wang, S.; Guo, J.; Yu, W.W.; Rogach, A.L. Metal Halide Perovskite Light-Emitting Devices: Promising Technology for Next-Generation Displays. *Adv. Funct. Mater.* **2019**, *29*, 1902008. [CrossRef]
6. Hassan, Y.; Park, J.H.; Crawford, M.L.; Sadhanala, A.; Lee, J.; Sadighian, J.C.; Mosconi, E.; Shivanna, R.; Radicchi, E.; Jeong, M.; et al. Ligand-Engineered Bandgap Stability in Mixed-Halide Perovskite LEDs. *Nature* **2021**, *591*, 72–77. [CrossRef] [PubMed]
7. Tan, Z.K.; Moghaddam, R.S.; Lai, M.L.; Docampo, P.; Higler, R.; Deschler, F.; Price, M.; Sadhanala, A.; Pazos, L.M.; Credgington, D.; et al. Bright Light-Emitting Diodes Based on Organometal Halide Perovskite. *Nat. Nanotechnol.* **2014**, *9*, 687–692. [CrossRef]
8. Zhou, W.; Shen, Y.; Cao, L.X.; Lu, Y.; Tang, Y.Y.; Zhang, K.; Ren, H.; Xie, F.M.; Li, Y.Q.; Tang, J.X. Manipulating Ionic Behavior with Bifunctional Additives for Efficient Sky-Blue Perovskite Light-Emitting Diodes. *Adv. Funct. Mater.* **2023**, *33*, 2301425. [CrossRef]
9. Bai, W.; Xuan, T.; Zhao, H.; Dong, H.; Cheng, X.; Wang, L.; Xie, R.J. Perovskite Light-Emitting Diodes with an External Quantum Efficiency Exceeding 30%. *Adv. Mater.* **2023**, *35*, 2302283. [CrossRef]
10. Jiang, J.; Chu, Z.; Yin, Z.; Li, J.; Yang, Y.; Chen, J.; Wu, J.; You, J.; Zhang, X. Red Perovskite Light-Emitting Diodes with Efficiency Exceeding 25% Realized by Co-Spacer Cations. *Adv. Mater.* **2022**, *34*, 2204460. [CrossRef]
11. Sun, Y.; Ge, L.; Dai, L.; Cho, C.; Ferrer Orri, J.; Ji, K.; Zelewski, S.J.; Liu, Y.; Mirabelli, A.J.; Zhang, Y.; et al. Bright and Stable Perovskite Light-Emitting Diodes in the near-Infrared Range. *Nature* **2023**, *615*, 830–835. [CrossRef] [PubMed]
12. Fakharuddin, A.; Gangishetty, M.K.; Abdi-Jalebi, M.; Chin, S.H.; bin Mohd Yusoff, A.R.; Congreve, D.N.; Tress, W.; Deschler, F.; Vasilopoulou, M.; Bolink, H.J. Perovskite Light-Emitting Diodes. *Nat. Electron.* **2022**, *5*, 203–216. [CrossRef]
13. Liu, X.K.; Xu, W.; Bai, S.; Jin, Y.; Wang, J.; Friend, R.H.; Gao, F. Metal Halide Perovskites for Light-Emitting Diodes. *Nat. Mater.* **2021**, *20*, 10–21. [CrossRef] [PubMed]
14. Tong, Y.; Bi, X.; Xu, S.; Min, H.; Cheng, L.; Kuang, Z.; Yuan, L.; Zhou, F.; Chu, Y.; Xu, L.; et al. In Situ Halide Exchange of Cesium Lead Halide Perovskites for Blue Light-Emitting Diodes. *Adv. Mater.* **2023**, *35*, 2207111. [CrossRef] [PubMed]
15. Xiong, W.; Zou, C.; Tang, W.; Xing, S.; Wang, Z.; Zhao, B.; Di, D. Efficient and Bright Blue Perovskite LEDs Enabled by a Carbazole-Phosphonic Acid Interface. *ACS Energy Lett.* **2023**, *8*, 2897–2903. [CrossRef]
16. Jang, C.H.; Kim, Y.I.; Harit, A.K.; Ha, J.M.; Park, S.; Noh, Y.W.; Lee, A.Y.; Kim, K.S.; Jung, J.W.; Woo, H.Y.; et al. Multifunctional Conjugated Molecular Additives for Highly Efficient Perovskite Light-Emitting Diodes. *Adv. Mater.* **2023**, *35*, 2210511. [CrossRef] [PubMed]
17. Zhao, Y.; Feng, W.; Li, M.; Lu, J.; Qin, X.; Lin, K.; Luo, J.; Zhang, W.H.; Lim, E.L.; Wei, Z. Efficient Perovskite Light-Emitting Diodes with Chemically Bonded Contact and Regulated Charge Behavior. *Nano Lett.* **2023**, *23*, 8560–8567. [CrossRef]
18. Jia, Y.; Yu, H.; Zhou, Y.; Li, N.; Guo, Y.; Xie, F.; Qin, Z.; Lu, X.; Zhao, N. Excess Ion-Induced Efficiency Roll-Off in High-Efficiency Perovskite Light-Emitting Diodes. *ACS Appl. Mater. Interfaces* **2021**, *13*, 28546–28554. [CrossRef]
19. Chu, Z.; Zhang, W.; Jiang, J.; Qu, Z.; Ma, F.; Zhao, Y.; Chu, X.; Shen, Y.; Li, Y.; Yin, Z.; et al. Blue Light-Emitting Diodes Based on Quasi-Two-Dimensional Perovskite with Efficient Charge Injection and Optimized Phase Distribution via an Alkali Metal Salt. *Nat. Electron.* **2023**, *6*, 360–369. [CrossRef]

20. Liu, S.; Guo, Z.; Wu, X.; Liu, X.; Huang, Z.; Li, L.; Zhang, J.; Zhou, H.; Sun, L.D.; Yan, C.H. Zwitterions Narrow Distribution of Perovskite Quantum Wells for Blue Light-Emitting Diodes with Efficiency Exceeding 15%. *Adv. Mater.* **2023**, *35*, 2208078. [CrossRef]
21. Ma, D.; Lin, K.; Dong, Y.; Choubisa, H.; Proppe, A.H.; Wu, D.; Wang, Y.K.; Chen, B.; Li, P.; Fan, J.Z.; et al. Distribution Control Enables Efficient Reduced-Dimensional Perovskite LEDs. *Nature* **2021**, *599*, 594–598. [CrossRef] [PubMed]
22. Lin, Y.K.; Chen, C.H.; Wang, Y.Y.; Yu, M.H.; Yang, J.W.; Ni, I.C.; Lin, B.H.; Zhidkov, I.S.; Kurmaev, E.Z.; Lu, Y.J.; et al. Realizing High Brightness Quasi-2D Perovskite Light-Emitting Diodes with Reduced Efficiency Roll-Off via Multifunctional Interface Engineering. *Adv. Sci.* **2023**, *10*, 2302232. [CrossRef] [PubMed]
23. Hong, Y.; Yu, C.; Je, H.; Park, J.Y.; Kim, T.; Baik, H.; Tomboc, G.M.; Kim, Y.; Ha, J.M.; Joo, J.; et al. Perovskite Nanocrystals Protected by Hermetically Sealing for Highly Bright and Stable Deep-Blue Light-Emitting Diodes. *Adv. Sci.* **2023**, *10*, 2302906. [CrossRef] [PubMed]
24. Jiang, Y.; Sun, C.; Xu, J.; Li, S.; Cui, M.; Fu, X.; Liu, Y.; Liu, Y.; Wan, H.; Wei, K.; et al. Synthesis-on-Substrate of Quantum Dot Solids. *Nature* **2022**, *612*, 679–684. [CrossRef] [PubMed]
25. Wan, Q.; Zheng, W.; Zou, C.; Carulli, F.; Zhang, C.; Song, H.; Liu, M.; Zhang, Q.; Lin, L.Y.; Kong, L.; et al. Ultrathin Light-Emitting Diodes with External Efficiency over 26% Based on Resurfaced Perovskite Nanocrystals. *ACS Energy Lett.* **2023**, *8*, 927–934. [CrossRef]
26. Mei, X.; Wang, J.; Zhang, X.; Zhuang, R.; Hua, Y.; He, K.; Zheng, W.; Zhang, X. In Situ Ligand Compensation of Perovskite Quantum Dots for Efficient Light-Emitting Diodes. *ACS Energy Lett.* **2023**, *8*, 4386–4396. [CrossRef]
27. Kim, J.S.; Heo, J.M.; Park, G.S.; Woo, S.J.; Cho, C.; Yun, H.J.; Kim, D.H.; Park, J.; Lee, S.C.; Park, S.H.; et al. Ultra-Bright, Efficient and Stable Perovskite Light-Emitting Diodes. *Nature* **2022**, *611*, 688–694. [CrossRef] [PubMed]
28. Zhao, F.; Duan, H.W.; Li, S.N.; Pan, J.L.; Shen, W.S.; Li, S.M.; Zhang, Q.; Wang, Y.K.; Liao, L.S. Iodotrimethylsilane as a Reactive Ligand for Surface Etching and Passivation of Perovskite Nanocrystals toward Efficient Pure-Red to Deep-Red LEDs. *Angew. Chem. Int. Ed.* **2023**, *62*, e202311089. [CrossRef]
29. Min, H.; Chang, J.; Tong, Y.; Wang, J.; Zhang, F.; Feng, Z.; Bi, X.; Chen, N.; Kuang, Z.; Wang, S.; et al. Additive Treatment Yields High-Performance Lead-Free Perovskite Light-Emitting Diodes. *Nat. Photonics* **2023**, *17*, 755–760. [CrossRef]
30. Han, D.; Wang, J.; Agosta, L.; Zang, Z.; Zhao, B.; Kong, L.; Lu, H.; Mosquera-Lois, I.; Carnevali, V.; Dong, J.; et al. Tautomeric Mixture Coordination Enables Efficient Lead-Free Perovskite LEDs. *Nature* **2023**, *622*, 493–498. [CrossRef]
31. Cao, L.; Klimes, K.; Ji, Y.; Fleetham, T.; Li, J. Efficient and Stable Organic Light-Emitting Devices Employing Phosphorescent Molecular Aggregates. *Nat. Photonics* **2021**, *15*, 230–237. [CrossRef]
32. Lee, T.; Kim, B.J.; Lee, H.; Hahm, D.; Bae, W.K.; Lim, J.; Kwak, J. Bright and Stable Quantum Dot Light-Emitting Diodes. *Adv. Mater.* **2022**, *34*, 2106276. [CrossRef] [PubMed]
33. Cao, Y.; Wang, N.; Tian, H.; Guo, J.; Wei, Y.; Chen, H.; Miao, Y.; Zou, W.; Pan, K.; He, Y.; et al. Perovskite Light-Emitting Diodes Based on Spontaneously Formed Submicrometre-Scale Structures. *Nature* **2018**, *562*, 249–253. [CrossRef] [PubMed]
34. Xiao, Z.; Kerner, R.A.; Zhao, L.; Tran, N.L.; Lee, K.M.; Koh, T.W.; Scholes, G.D.; Rand, B.P. Efficient Perovskite Light-Emitting Diodes Featuring Nanometre-Sized Crystallites. *Nat. Photonics* **2017**, *11*, 108–115. [CrossRef]
35. Zhao, X.; Tan, Z.K. Large-Area near-Infrared Perovskite Light-Emitting Diodes. *Nat. Photonics* **2020**, *14*, 215–218. [CrossRef]
36. Zhao, B.; Lian, Y.; Cui, L.; Divitini, G.; Kusch, G.; Ruggeri, E.; Auras, F.; Li, W.; Yang, D.; Zhu, B.; et al. Efficient Light-Emitting Diodes from Mixed-Dimensional Perovskites on a Fluoride Interface. *Nat. Electron.* **2020**, *3*, 704–710. [CrossRef]
37. Zhao, B.; Bai, S.; Kim, V.; Lamboll, R.; Shivanna, R.; Auras, F.; Richter, J.M.; Yang, L.; Dai, L.; Alsari, M.; et al. High-Efficiency Perovskite–Polymer Bulk Heterostructure Light-Emitting Diodes. *Nat. Photonics* **2018**, *12*, 783–789. [CrossRef]
38. Zhao, B.; Vasilopoulou, M.; Fakharuddin, A.; Gao, F.; Mohd Yusoff, A.R.B.; Friend, R.H.; Di, D. Light Management for Perovskite Light-Emitting Diodes. *Nat. Nanotechnol.* **2023**, *18*, 981–992. [CrossRef]
39. Moore, D.T.; Sai, H.; Tan, K.W.; Smilgies, D.M.; Zhang, W.; Snaith, H.J.; Wiesner, U.; Estroff, L.A. Crystallization Kinetics of Organic-Inorganic Trihalide Perovskites and the Role of the Lead Anion in Crystal Growth. *J. Am. Chem. Soc.* **2015**, *137*, 2350–2358. [CrossRef]
40. Chu, Z.; Yang, M.; Schulz, P.; Wu, D.; Ma, X.; Seifert, E.; Sun, L.; Li, X.; Zhu, K.; Lai, K. Impact of Grain Boundaries on Efficiency and Stability of Organic-Inorganic Trihalide Perovskites. *Nat. Commun.* **2017**, *8*, 2230. [CrossRef]
41. Nie, W.; Tsai, H.; Asadpour, R.; Neukirch, A.J.; Gupta, G.; Crochet, J.J.; Chhowalla, M.; Tretiak, S.; Alam, M.A.; Wang, H.; et al. High-Efficiency Solution-Processed Perovskite Solar Cells with Millimeter-Scale Grains. *Science* **2015**, *347*, 522–525. [CrossRef] [PubMed]
42. Jeong, J.; Kim, M.; Seo, J.; Lu, H.; Ahlawat, P.; Mishra, A.; Yang, Y.; Hope, M.A.; Eickemeyer, F.T.; Kim, M.; et al. Pseudo-Halide Anion Engineering for α-FAPbI3 Perovskite Solar Cells. *Nature* **2021**, *592*, 381–385. [CrossRef] [PubMed]
43. Zhao, L.; Tang, P.; Luo, D.; Dar, M.I.; Eickemeyer, F.T.; Arora, N.; Hu, Q.; Luo, J.; Liu, Y.; Zakeeruddin, S.M.; et al. Enabling Full-Scale Grain Boundary Mitigation in Polycrystalline Perovskite Solids. *Sci. Adv.* **2022**, *8*, eabo3733. [CrossRef] [PubMed]
44. Di Girolamo, D.; Dar, M.I.; Dini, D.; Gontrani, L.; Caminiti, R.; Mattoni, A.; Graetzel, M.; Meloni, S. Dual Effect of Humidity on Cesium Lead Bromide: Enhancement and Degradation of Perovskite Films. *J. Mater. Chem. A* **2019**, *7*, 12292–12302. [CrossRef]
45. Wang, S.; Jiang, Y.; Juarez-Perez, E.J.; Ono, L.K.; Qi, Y. Accelerated Degradation of Methylammonium Lead Iodide Perovskites Induced by Exposure to Iodine Vapour. *Nat. Energy* **2017**, *2*, 16195. [CrossRef]

46. Ghosh, S.; Pal, S.K.; Karki, K.J.; Pullerits, T. Ion Migration Heals Trapping Centers in CH3NH3PbBr3 Perovskite. *ACS Energy Let* **2017**, *2*, 2133–2139. [CrossRef]
47. Aristidou, N.; Eames, C.; Sanchez-Molina, I.; Bu, X.; Kosco, J.; Saiful Islam, M.; Haque, S.A. Fast Oxygen Diffusion and Iodid Defects Mediate Oxygen-Induced Degradation of Perovskite Solar Cells. *Nat. Commun.* **2017**, *8*, 15218. [CrossRef]
48. Han, T.H.; Lee, J.W.; Choi, Y.J.; Choi, C.; Tan, S.; Lee, S.J.; Zhao, Y.; Huang, Y.; Kim, D.; Yang, Y. Surface-2D/Bulk-3D Heterophase Perovskite Nanograins for Long-Term-Stable Light-Emitting Diodes. *Adv. Mater.* **2020**, *32*, 1905674. [CrossRef]
49. Vashishtha, P.; Halpert, J.E. Field-Driven Ion Migration and Color Instability in Red-Emitting Mixed Halide Perovskite Nanocryst Light-Emitting Diodes. *Chem. Mater.* **2017**, *29*, 5965–5973. [CrossRef]
50. Fan, Z.; Xiao, H.; Wang, Y.; Zhao, Z.; Lin, Z.; Cheng, H.C.; Lee, S.J.; Wang, G.; Feng, Z.; Goddard, W.A.; et al. Layer-by-Laye Degradation of Methylammonium Lead Tri-Iodide Perovskite Microplates. *Joule* **2017**, *1*, 548–562. [CrossRef]
51. Eperon, G.E.; Habisreutinger, S.N.; Leijtens, T.; Bruijnaers, B.J.; Van Franeker, J.J.; Dequilettes, D.W.; Pathak, S.; Sutton, R.J Grancini, G.; Ginger, D.S.; et al. The Importance of Moisture in Hybrid Lead Halide Perovskite Thin Film Fabrication. *ACS Nan* **2015**, *9*, 9380–9393. [CrossRef] [PubMed]
52. Rehman, W.; McMeekin, D.P.; Patel, J.B.; Milot, R.L.; Johnston, M.B.; Snaith, H.J.; Herz, L.M. Photovoltaic Mixed-Cation Lea Mixed-Halide Perovskites: Links between Crystallinity, Photo-Stability and Electronic Properties. *Energy Environ. Sci.* **2017**, *1* 361–369. [CrossRef]
53. Zhang, L.; Yang, X.; Jiang, Q.; Wang, P.; Yin, Z.; Zhang, X.; Tan, H.; Yang, Y.M.; Wei, M.; Sutherland, B.R.; et al. Ultra-Bright an Highly Efficient Inorganic Based Perovskite Light-Emitting Diodes. *Nat. Commun.* **2017**, *8*, 15640. [CrossRef] [PubMed]
54. Jiang, Y.; Qin, C.; Cui, M.; He, T.; Liu, K.; Huang, Y.; Luo, M.; Zhang, L.; Xu, H.; Li, S.; et al. Spectra Stable Blue Perovskit Light-Emitting Diodes. *Nat. Commun.* **2019**, *10*, 1868. [CrossRef] [PubMed]
55. Cao, J.; Tao, S.X.; Bobbert, P.A.; Wong, C.P.; Zhao, N. Interstitial Occupancy by Extrinsic Alkali Cations in Perovskites and It Impact on Ion Migration. *Adv. Mater.* **2018**, *30*, 1707350. [CrossRef] [PubMed]
56. Du, P.; Li, J.; Wang, L.; Sun, L.; Wang, X.; Xu, X.; Yang, L.; Pang, J.; Liang, W.; Luo, J.; et al. Efficient and Large-Area A Vacuum-Deposited Perovskite Light-Emitting Diodes via Spatial Confinement. *Nat. Commun.* **2021**, *12*, 4751. [CrossRef [PubMed]
57. Cho, H.; Kim, Y.H.; Wolf, C.; Lee, H.D.; Lee, T.W. Improving the Stability of Metal Halide Perovskite Materials and Light-Emittin Diodes. *Adv. Mater.* **2018**, *30*, 1704587. [CrossRef]
58. Liu, M.; Matuhina, A.; Zhang, H.; Vivo, P. Advances in the Stability of Halide Perovskite Nanocrystals. *Materials* **2019**, *12*, 373 [CrossRef]
59. Dong, Y.; Wang, Y.K.; Yuan, F.; Johnston, A.; Liu, Y.; Ma, D.; Choi, M.J.; Chen, B.; Chekini, M.; Baek, S.W.; et al. Bipolar-Shel Resurfacing for Blue LEDs Based on Strongly Confined Perovskite Quantum Dots. *Nat. Nanotechnol.* **2020**, *15*, 668–674. [CrossRef
60. Mei, A.; Li, X.; Liu, L.; Ku, Z.; Liu, T.; Rong, Y.; Xu, M.; Hu, M.; Chen, J.; Yang, Y.; et al. A Hole-Conductor-Free, Fully Printabl Mesoscopic Perovskite Solar Cell with High Stability. *Science* **2014**, *345*, 295–298. [CrossRef]
61. Han, G.S.; Kim, J.; Bae, S.; Han, S.; Kim, Y.J.; Gong, O.Y.; Lee, P.; Ko, M.J.; Jung, H.S. Spin-Coating Process for 10 Cm × 10 Cm Perovskite Solar Modules Enabled by Self-Assembly of SnO_2 Nanocolloids. *ACS Energy Lett.* **2019**, *4*, 1845–1851. [CrossRef]
62. Jeong, D.N.; Lee, D.K.; Seo, S.; Lim, S.Y.; Zhang, Y.; Shin, H.; Cheong, H.; Park, N.G. Perovskite Cluster-Containing Solution fo Scalable D-Bar Coating toward High-Throughput Perovskite Solar Cells. *ACS Energy Lett.* **2019**, *4*, 1189–1195. [CrossRef]
63. Xu, M.; Ji, W.; Sheng, Y.; Wu, Y.; Cheng, H.; Meng, J.; Yan, Z.; Xu, J.; Mei, A.; Hu, Y.; et al. Efficient Triple-Mesoscopic Perovskite Solar Mini-Modules Fabricated with Slot-Die Coating. *Nano Energy* **2020**, *74*, 104842. [CrossRef]
64. Zeng, L.; Chen, S.; Forberich, K.; Brabec, C.J.; Mai, Y.; Guo, F. Controlling the Crystallization Dynamics of Photovoltaic Perovskite Layers on Larger-Area Coatings. *Energy Environ. Sci.* **2020**, *13*, 4666–4690. [CrossRef]
65. Ball, J.M.; Petrozza, A. Defects in Perovskite-Halides and Their Effects in Solar Cells. *Nat. Energy* **2016**, *1*, 16149. [CrossRef]
66. Jeong, J.E.; Park, J.H.; Jang, C.H.; Song, M.H.; Woo, H.Y. Multifunctional Charge Transporting Materials for Perovskite Light Emitting Diodes. *Adv. Mater.* **2020**, *32*, 2002176. [CrossRef]
67. Cao, J.; Yin, J.; Yuan, S.; Zhao, Y.; Li, J.; Zheng, N. Thiols as Interfacial Modifiers to Enhance the Performance and Stability of Perovskite Solar Cells. *Nanoscale* **2015**, *7*, 9443–9447. [CrossRef]
68. Wang, R.; Xue, J.; Meng, L.; Lee, J.W.; Zhao, Z.; Sun, P.; Cai, L.; Huang, T.; Wang, Z.; Wang, Z.K.; et al. Caffeine Improves the Performance and Thermal Stability of Perovskite Solar Cells. *Joule* **2019**, *3*, 1464–1477. [CrossRef]
69. Huang, G.; Wang, C.; Zhang, H.; Xu, S.; Xu, Q.; Cui, Y. Post-Healing of Defects: An Alternative Way for Passivation of Carbon-Based Mesoscopic Perovskite Solar Cells: Via Hydrophobic Ligand Coordination. *J. Mater. Chem. A* **2018**, *6*, 2449–2455 [CrossRef]
70. Du, K.; Wang, A.; Li, Y.; Xu, Y.; Li, L.; Yuan, N.; Ding, J. The Synergistic Effect of Phosphonic and Carboxyl Acid Groups for Efficient and Stable Perovskite Solar Cells. *Materials* **2023**, *16*, 7306. [CrossRef]
71. Wang, K.L.; Li, M.; Lou, Y.H.; Chen, J.; Shi, Y.R.; Chen, C.H.; Zhou, Y.H.; Wang, Z.K.; Liao, L.S. Aniline Sulfonic Acid Induced Uniform Perovskite Film for Large-Scale Photovoltaics. *Adv. Energy Mater.* **2023**, *13*, 2203471. [CrossRef]
72. Zhang, H.; Pfeifer, L.; Zakeeruddin, S.M.; Chu, J.; Grätzel, M. Tailoring Passivators for Highly Efficient and Stable Perovskite Solar Cells. *Nat. Rev. Chem.* **2023**, *7*, 632–652. [CrossRef]
73. Jiang, Q.; Tong, J.; Xian, Y.; Kerner, R.A.; Dunfield, S.P.; Xiao, C.; Scheidt, R.A.; Kuciauskas, D.; Wang, X.; Hautzinger, M.P.; et al. Surface Reaction for Efficient and Stable Inverted Perovskite Solar Cells. *Nature* **2022**, *611*, 278–283. [CrossRef]

74. Li, G.; Tan, Z.K.; Di, D.; Lai, M.L.; Jiang, L.; Lim, J.H.W.; Friend, R.H.; Greenham, N.C. Efficient Light-Emitting Diodes Based on Nanocrystalline Perovskite in a Dielectric Polymer Matrix. *Nano Lett.* **2015**, *15*, 2640–2644. [CrossRef] [PubMed]
75. Ji, X.; Peng, X.; Wang, Q.; Ren, J.; Xiong, Z.; Yang, X. On the Performance of Polymer:Organometal Halide Perovskite Composite Light Emitting Devices: The Effects of Polymer Additives. *Org. Electron.* **2018**, *52*, 350–355. [CrossRef]
76. Kumawat, N.K.; Dey, A.; Kumar, A.; Gopinathan, S.P.; Narasimhan, K.L.; Kabra, D. Band Gap Tuning of CH3NH3Pb(Br1-XClx)3 Hybrid Perovskite for Blue Electroluminescence. *ACS Appl. Mater. Interfaces* **2015**, *7*, 13119–13124. [CrossRef] [PubMed]
77. Chen, Z.; Zhang, C.; Jiang, X.F.; Liu, M.; Xia, R.; Shi, T.; Chen, D.; Xue, Q.; Zhao, Y.J.; Su, S.; et al. High-Performance Color-Tunable Perovskite Light Emitting Devices through Structural Modulation from Bulk to Layered Film. *Adv. Mater.* **2017**, *29*, 1603157. [CrossRef]
78. Yoon, Y.J.; Shin, Y.S.; Jang, H.; Son, J.G.; Kim, J.W.; Park, C.B.; Yuk, D.; Seo, J.; Kim, G.H.; Kim, J.Y. Highly Stable Bulk Perovskite for Blue LEDs with Anion-Exchange Method. *Nano Lett.* **2021**, *21*, 3473–3479. [CrossRef]
79. Park, C.B.; Shin, Y.S.; Yoon, Y.J.; Jang, H.; Son, J.G.; Kim, S.; An, N.G.; Kim, J.W.; Jun, Y.C.; Kim, G.H.; et al. Suppression of Halide Migration and Immobile Ionic Surface Passivation for Blue Perovskite Light-Emitting Diodes. *J. Mater. Chem. C* **2022**, *10*, 2060–2066. [CrossRef]
80. Cheng, L.; Yi, C.; Tong, Y.; Zhu, L.; Kusch, G.; Wang, X.; Wang, X.; Jiang, T.; Zhang, H.; Zhang, J.; et al. Halide Homogenization for High-Performance Blue Perovskite Electroluminescence. *Research* **2020**, *2020*, 9017871. [CrossRef]
81. Li, L.; Hu, Y.; Chen, Y.; Wang, C.; Zhao, G.; Du, X.; Wang, C.; Xiao, L.; Lu, Z.; Wang, J.; et al. Surface Defect Suppression for High Color Purity Light-Emitting Diode of Free-Standing Single-Crystal Perovskite Film. *Adv. Funct. Mater.* **2023**, *33*, 2301205. [CrossRef]
82. Wang, Q.; Chen, Y.; Yan, C.; Zeng, X.; Fu, X.; Pan, L.; Cao, J.; Yang, S.; Li, W.; Chen, X.; et al. Molecularly Designing a Passivation ETL to Suppress EQE Roll-off of PeLEDs. *ACS Energy Lett.* **2023**, *8*, 3710–3719. [CrossRef]
83. Wei, J.; Li, J.; Duan, C.; Yuan, L.; Zou, S.; Pang, Q.; Yan, K. High Efficiency Near-Infrared Perovskite Light Emitting Diodes With Reduced Rolling-Off by Surface Post-Treatment. *Small* **2023**, *19*, e2207769. [CrossRef]
84. Lin, K.; Xing, J.; Quan, L.N.; de Arquer, F.P.G.; Gong, X.; Lu, J.; Xie, L.; Zhao, W.; Zhang, D.; Yan, C.; et al. Perovskite Light-Emitting Diodes with External Quantum Efficiency Exceeding 20 per Cent. *Nature* **2018**, *562*, 245–248. [CrossRef] [PubMed]
85. Li, Z.; Sun, X.; Zheng, X.; Li, B.; Gao, D.; Zhang, S.; Wu, X.; Li, S.; Gong, J.; Luther, J.M.; et al. Stabilized Hole-Selective Layer for High-Performance Inverted p-i-n Perovskite Solar Cells. *Science* **2023**, *382*, 284–289. [CrossRef]
86. Zhang, S.; Ye, F.; Wang, X.; Chen, R.; Zhang, H.; Zhan, L.; Jiang, X.; Li, Y.; Ji, X.; Liu, S.; et al. Minimizing Buried Interfacial Defects for Efficient Inverted Perovskite Solar Cells. *Science* **2023**, *380*, 404–409. [CrossRef] [PubMed]
87. Chang, C.Y.; Chu, C.Y.; Huang, Y.C.; Huang, C.W.; Chang, S.Y.; Chen, C.A.; Chao, C.Y.; Su, W.F. Tuning Perovskite Morphology by Polymer Additive for High Efficiency Solar Cell. *ACS Appl. Mater. Interfaces* **2015**, *7*, 4955–4961. [CrossRef]
88. Wei, J.; Huang, F.; Wang, S.; Zhou, L.; Jin, P.; Xin, Y.; Cai, Z.; Yin, Z.; Pang, Q.; Zhang, J.Z. Highly Stable Hybrid Perovskite Solar Cells Modified with Polyethylenimine via Ionic Bonding. *ChemNanoMat* **2018**, *4*, 649–655. [CrossRef]
89. Kim, M.; Motti, S.G.; Sorrentino, R.; Petrozza, A. Enhanced Solar Cell Stability by Hygroscopic Polymer Passivation of Metal Halide Perovskite Thin Film. *Energy Environ. Sci.* **2018**, *11*, 2609–2619. [CrossRef]
90. Shivarudraiah, S.B.; Chan, C.C.S.; Chen, D.; Zhou, Z.; Ng, M.; Tewari, N.; Tao, C.K.; Wong, K.S.; Halpert, J.E. In Situ MAPbI3 Perovskite Nanostructures Formed by a Poly[(Phenylglycidyl Ether)-Co-Formaldehyde] Epoxide for Enhanced Stability and Photoluminescence. *ACS Appl. Nano Mater.* **2023**, *6*, 12240–12247. [CrossRef]
91. Calabrese, J.; Jones, N.L.; Harlow, R.L.; Herron, N.D.; Thorn, L.; Wang, Y. Preparation and Characterization of Layered Rare Earth Compound. *J. Am. Chem. Soc.* **1991**, *113*, 2328–2330. [CrossRef]
92. Hanmandlu, C.; Singh, A.; Chen, L.; Wang, H.; Fang, C.; Luo, H. Two/Quasi-Two-Dimensional Perovskite-Based Heterostructures: Construction, Properties and Applications. *Int. J. Extrem. Manuf.* **2023**, *5*, 012004.
93. Stoumpos, C.C.; Cao, D.H.; Clark, D.J.; Young, J.; Rondinelli, J.M.; Jang, J.I.; Hupp, J.T.; Kanatzidis, M.G. Ruddlesden-Popper Hybrid Lead Iodide Perovskite 2D Homologous Semiconductors. *Chem. Mater.* **2016**, *28*, 2852–2867. [CrossRef]
94. Lee, J.-W.; Park, N.-G. Quasi-Two-Dimensional Perovskite Light Emitting Diodes for Bright Future. *Light Sci. Appl.* **2021**, *10*, 10–11. [CrossRef] [PubMed]
95. Chen, Y.; Sun, Y.; Peng, J.; Tang, J.; Zheng, K.; Liang, Z. 2D Ruddlesden–Popper Perovskites for Optoelectronics. *Adv. Mater.* **2018**, *30*, 1703487. [CrossRef]
96. Yuan, M.; Quan, L.N.; Comin, R.; Walters, G.; Sabatini, R.; Voznyy, O.; Hoogland, S.; Zhao, Y.; Beauregard, E.M.; Kanjanaboos, P.; et al. Perovskite Energy Funnels for Efficient Light-Emitting Diodes. *Nat. Nanotechnol.* **2016**, *11*, 872–877. [CrossRef] [PubMed]
97. Tsai, H.; Nie, W.; Blancon, J.C.; Stoumpos, C.C.; Asadpour, R.; Harutyunyan, B.; Neukirch, A.J.; Verduzco, R.; Crochet, J.J.; Tretiak, S.; et al. High-Efficiency Two-Dimensional Ruddlesden-Popper Perovskite Solar Cells. *Nature* **2016**, *536*, 312–317. [CrossRef]
98. Wu, C.; Wu, T.; Yang, Y.; McLeod, J.A.; Wang, Y.; Zou, Y.; Zhai, T.; Li, J.; Ban, M.; Song, T.; et al. Alternative Type Two-Dimensional-Three-Dimensional Lead Halide Perovskite with Inorganic Sodium Ions as a Spacer for High-Performance Light-Emitting Diodes. *ACS Nano* **2019**, *13*, 1645–1654. [CrossRef]
99. Cheng, L.; Jiang, T.; Cao, Y.; Yi, C.; Wang, N.; Huang, W.; Wang, J. Multiple-Quantum-Well Perovskites for High-Performance Light-Emitting Diodes. *Adv. Mater.* **2020**, *32*, 201904163. [CrossRef]

100. Quan, L.N.; Zhao, Y.; García De Arquer, F.P.; Sabatini, R.; Walters, G.; Voznyy, O.; Comin, R.; Li, Y.; Fan, J.Z.; Tan, H.; et al Tailoring the Energy Landscape in Quasi-2D Halide Perovskites Enables Efficient Green-Light Emission. *Nano Lett.* **2017**, *17*, 3701–3709. [CrossRef]
101. Kong, L.; Zhang, X.; Li, Y.; Wang, H.; Jiang, Y.; Wang, S.; You, M.; Zhang, C.; Zhang, T.; Kershaw, S.V.; et al. Smoothing the Energy Transfer Pathway in Quasi-2D Perovskite Films Using Methanesulfonate Leads to Highly Efficient Light-Emitting Devices. *Nat. Commun.* **2021**, *12*, 1246. [CrossRef] [PubMed]
102. Jiang, Y.; Cui, M.; Li, S.; Sun, C.; Huang, Y.; Wei, J.; Zhang, L.; Lv, M.; Qin, C.; Liu, Y.; et al. Reducing the Impact of Auger Recombination in Quasi-2D Perovskite Light-Emitting Diodes. *Nat. Commun.* **2021**, *12*, 336. [CrossRef] [PubMed]
103. Yang, S.J.; Wang, K.; Luo, Y.; Park, J.Y.; Yang, H.; Coffey, A.H.; Ma, K.; Sun, J.; Wieghold, S.; Zhu, C.; et al. Two-Factor Phase Separations in Mixed-Halide Quasi-2D Perovskite LEDs: Dimensionality and Halide Segregations. *ACS Energy Lett.* **2023**, *8*, 3693–3701. [CrossRef]
104. Zhong, J.X.; Wu, W.Q.; Liao, J.F.; Feng, W.; Jiang, Y.; Wang, L.; Kuang, D. Bin The Rise of Textured Perovskite Morphology: Revolutionizing the Pathway toward High-Performance Optoelectronic Devices. *Adv. Energy Mater.* **2020**, *10*, 1902256. [CrossRef]
105. Chen, Z.; Li, Z.; Chen, Z.; Xia, R.; Zou, G.; Chu, L.; Su, S.J.; Peng, J.; Yip, H.L.; Cao, Y. Utilization of Trapped Optical Modes for White Perovskite Light-Emitting Diodes with Efficiency over 12%. *Joule* **2021**, *5*, 456–466. [CrossRef]
106. Cui, J.; Liu, Y.; Deng, Y.; Lin, C.; Fang, Z.; Xiang, C.; Bai, P.; Du, K.; Zuo, X.; Wen, K.; et al. Efficient Light-Emitting Diodes Based on Oriented Perovskite Nanoplatelets. *Sci. Adv.* **2021**, *7*, eabg8458. [CrossRef] [PubMed]
107. Kumar, A.; Singh, S.; Yadav, A. Recent Progress in Inverted Perovskite Solar Cells Employing Nickel Oxide (NiOx) as a Hole Transport Materials. *Mater. Today Proc.* **2021**, *46*, 5827–5832. [CrossRef]
108. Kazes, M.; Udayabhaskararao, T.; Dey, S.; Oron, D. Effect of Surface Ligands in Perovskite Nanocrystals: Extending in and Reaching Out. *Acc. Chem. Res.* **2021**, *54*, 1409–1418. [CrossRef]
109. Kovalenko, M.V.; Protesescu, L.; Bodnarchuk, M.I. Properties and Potential Optoelectronic Applications of Lead Halide Perovskite Nanocrystals. *Science* **2017**, *358*, 745–750. [CrossRef]
110. Akkerman, Q.A.; Rainò, G.; Kovalenko, M.V.; Manna, L. Genesis, Challenges and Opportunities for Colloidal Lead Halide Perovskite Nanocrystals. *Nat. Mater.* **2018**, *17*, 394–405. [CrossRef]
111. Shamsi, J.; Rainò, G.; Kovalenko, M.V.; Stranks, S.D. To Nano or Not to Nano for Bright Halide Perovskite Emitters. *Nat. Nanotechnol.* **2021**, *16*, 1164–1168. [CrossRef] [PubMed]
112. Schmidt, L.C.; Pertegás, A.; Gonzalez-carrero, S.; Malinkiewicz, O.; Agouram, S.; Espallargas, G.M.; Bolink, H.J.; Galian, R.E.; Pérez-prieto, J.; González-carrero, S. Non-Template Synthesis of CH3NH3PbBr3 Perovskite Nanoparticles Non-Template Synthesis of CH 3 NH 3 PbBr 3 Perovskite Nanoparticles. *J. Am. Chem. Soc.* **2014**, *136*, 850–853. [CrossRef] [PubMed]
113. Kim, Y.H.; Kim, S.; Kakekhani, A.; Park, J.; Park, J.; Lee, Y.H.; Xu, H.; Nagane, S.; Wexler, R.B.; Kim, D.H.; et al. Comprehensive Defect Suppression in Perovskite Nanocrystals for High-Efficiency Light-Emitting Diodes. *Nat. Photonics* **2021**, *15*, 148–155. [CrossRef]
114. Kim, Y.H.; Park, J.; Kim, S.; Kim, J.S.; Xu, H.; Jeong, S.H.; Hu, B.; Lee, T.W. Exploiting the Full Advantages of Colloidal Perovskite Nanocrystals for Large-Area Efficient Light-Emitting Diodes. *Nat. Nanotechnol.* **2022**, *17*, 590–597. [CrossRef] [PubMed]
115. Yun, Q.; Ge, Y.; Huang, B.; Wa, Q.; Zhang, H. Ligand-Assisted Phase Engineering of Nanomaterials. *Acc. Chem. Res.* **2023**, *56*, 1780–1790. [CrossRef]
116. Kim, Y.H.; Lee, T.W. Engineering Colloidal Perovskite Nanocrystals and Devices for Efficient and Large-Area Light-Emitting Diodes. *Accounts Mater. Res.* **2023**, *4*, 655–667. [CrossRef]
117. Yassitepe, E.; Yang, Z.; Voznyy, O.; Kim, Y.; Walters, G.; Castañeda, J.A.; Kanjanaboos, P.; Yuan, M.; Gong, X.; Fan, F.; et al. Amine-Free Synthesis of Cesium Lead Halide Perovskite Quantum Dots for Efficient Light-Emitting Diodes. *Adv. Funct. Mater.* **2016**, *26*, 8757–8763. [CrossRef]
118. Shynkarenko, Y.; Bodnarchuk, M.I.; Bernasconi, C.; Berezovska, Y.; Verteletskyi, V.; Ochsenbein, S.T.; Kovalenko, M.V. Direct Synthesis of Quaternary Alkylammonium-Capped Perovskite Nanocrystals for Efficient Blue and Green Light-Emitting Diodes. *ACS Energy Lett.* **2019**, *4*, 2703–2711. [CrossRef]
119. Yang, D.; Li, X.; Zhou, W.; Zhang, S.; Meng, C.; Wu, Y.; Wang, Y.; Zeng, H. CsPbBr3 Quantum Dots 2.0: Benzenesulfonic Acid Equivalent Ligand Awakens Complete Purification. *Adv. Mater.* **2019**, *31*, 1900767. [CrossRef]
120. Zhao, C.; Zhu, C.; Yu, Y.; Xue, W.; Liu, X.; Yuan, F.; Dai, J.; Wang, S.; Jiao, B.; Wu, Z. Multifunctional Short-Chain 2-Thiophenealkylammonium Bromide Ligand-Assisted Perovskite Quantum Dots for Efficient Light-Emitting Diodes. *ACS Appl. Mater. Interfaces* **2023**, *15*, 40080–40087. [CrossRef]
121. Guo, J.; Lu, M.; Zhang, X.; Sun, S.; Han, C.; Zhang, Y.; Yang, X.; Kershaw, S.V.; Zheng, W.; Rogach, A.L. Highly Stable and Efficient Light-Emitting Diodes Based on Orthorhombic γ-CsPbI3Nanocrystals. *ACS Nano* **2023**, *17*, 9290–9301. [CrossRef] [PubMed]
122. Pan, J.; Quan, L.N.; Zhao, Y.; Peng, W.; Murali, B.; Sarmah, S.P.; Yuan, M.; Sinatra, L.; Alyami, N.M.; Liu, J.; et al. Highly Efficient Perovskite-Quantum-Dot Light-Emitting Diodes by Surface Engineering. *Adv. Mater.* **2016**, *28*, 8718–8725. [CrossRef] [PubMed]
123. Boles, M.A.; Ling, D.; Hyeon, T.; Talapin, D.V. The Surface Science of Nanocrystals. *Nat. Mater.* **2016**, *15*, 141–153. [CrossRef] [PubMed]
124. Yin, J.; Yang, H.; Gutiérrez-Arzaluz, L.; Zhou, Y.; Brédas, J.L.; Bakr, O.M.; Mohammed, O.F. Luminescence and Stability Enhancement of Inorganic Perovskite Nanocrystals via Selective Surface Ligand Binding. *ACS Nano* **2021**, *15*, 17998–18005. [CrossRef] [PubMed]

25. Chen, D.; Ko, P.K.; Li, C.H.A.; Zou, B.; Geng, P.; Guo, L.; Halpert, J.E. Amino Acid-Passivated Pure Red CsPbI3 Quantum Dot LEDs. *ACS Energy Lett.* **2023**, *8*, 410–416. [CrossRef]
26. Fiuza-Maneiro, N.; Sun, K.; López-Fernández, I.; Gómez-Graña, S.; Müller-Buschbaum, P.; Polavarapu, L. Ligand Chemistry of Inorganic Lead Halide Perovskite Nanocrystals. *ACS Energy Lett.* **2023**, *8*, 1152–1191. [CrossRef]
27. Yoon, S.M.; Min, H.; Kim, J.B.; Kim, G.; Lee, K.S.; Seok, S. Il Surface Engineering of Ambient-Air-Processed Cesium Lead Triiodide Layers for Efficient Solar Cells. *Joule* **2021**, *5*, 183–196. [CrossRef]
28. Wang, Y.K.; Yuan, F.; Dong, Y.; Li, J.Y.; Johnston, A.; Chen, B.; Saidaminov, M.I.; Zhou, C.; Zheng, X.; Hou, Y.; et al. All-Inorganic Quantum-Dot LEDs Based on a Phase-Stabilized α-CsPbI3 Perovskite. *Angew. Chem. Int. Ed.* **2021**, *60*, 16164–16170. [CrossRef]
29. Wang, H.; Liu, H.; Dong, Z.; Wei, X.; Li, W.; Zhu, L.; Zhu, C.; Bai, Y.; Chen, H. Dimethyl Sulfoxide: A Promising Solvent for Inorganic CsPbI3 Perovskite. *Sci. Bull.* **2023**, *68*, 192–202. [CrossRef]
30. Koscher, B.A.; Swabeck, J.K.; Bronstein, N.D.; Alivisatos, A.P. Essentially Trap-Free CsPbBr3 Colloidal Nanocrystals by Postsynthetic Thiocyanate Surface Treatment. *J. Am. Chem. Soc.* **2017**, *139*, 6566–6569. [CrossRef]
31. Kim, Y.H.; Kim, J.S.; Lee, T.W. Strategies to Improve Luminescence Efficiency of Metal-Halide Perovskites and Light-Emitting Diodes. *Adv. Mater.* **2019**, *31*, 1804595. [CrossRef] [PubMed]
32. Yang, F.; Chen, H.; Zhang, R.; Liu, X.; Zhang, W.; Zhang, J.B.; Gao, F.; Wang, L. Efficient and Spectrally Stable Blue Perovskite Light-Emitting Diodes Based on Potassium Passivated Nanocrystals. *Adv. Funct. Mater.* **2020**, *30*, 1908760. [CrossRef]
33. Song, J.; Fang, T.; Li, J.; Xu, L.; Zhang, F.; Han, B.; Shan, Q.; Zeng, H. Organic–Inorganic Hybrid Passivation Enables Perovskite QLEDs with an EQE of 16.48%. *Adv. Mater.* **2018**, *30*, 1805409. [CrossRef] [PubMed]
34. Xu, J.; Yan, H.; Ali, M.U.; Wu, L.; Miao, J.; Cai, J.; Ning, J.; Yang, B.; Liu, M.; Shen, C.K.F.; et al. Efficient NIR Perovskite Light-Emitting Diodes Enabled by Incorporating an Anthracene Derivative as a Bifunctional Electron Transport Layer. *ACS Appl. Electron. Mater.* **2022**, *4*, 1669–1677. [CrossRef]
35. Chiba, T.; Hayashi, Y.; Ebe, H.; Hoshi, K.; Sato, J.; Sato, S.; Pu, Y.J.; Ohisa, S.; Kido, J. Anion-Exchange Red Perovskite Quantum Dots with Ammonium Iodine Salts for Highly Efficient Light-Emitting Devices. *Nat. Photonics* **2018**, *12*, 681–687. [CrossRef]
136. Yuan, S.; Fang, T.; Huang, J.; Li, X.; Wei, C.; Zhou, Y.; Li, Y.; Zheng, X.; Huang, J.; Su, J.; et al. Balancing Charge Injection via a Tailor-Made Electron-Transporting Material for High Performance Blue Perovskite QLEDs. *ACS Energy Lett.* **2023**, *8*, 818–826. [CrossRef]
137. Jiang, Y.; Wei, K.; Sun, C.; Feng, Y.; Zhang, L.; Cui, M.; Li, S.; Li, W.D.; Kim, J.T.; Qin, C.; et al. Unraveling Size-Dependent Ion-Migration for Stable Mixed-Halide Perovskite Light-Emitting Diodes. *Adv. Mater.* **2023**, *35*, 2304094. [CrossRef]
138. Wang, H.; Xu, W.; Wei, Q.; Peng, S.; Shang, Y.; Jiang, X.; Yu, D.; Wang, K.; Pu, R.; Zhao, C.; et al. In-Situ Growth of Low-Dimensional Perovskite-Based Insular Nanocrystals for Highly Efficient Light Emitting Diodes. *Light Sci. Appl.* **2023**, *12*, 62. [CrossRef]
139. López-Fernández, I.; Valli, D.; Wang, C.Y.; Samanta, S.; Okamoto, T.; Huang, Y.T.; Sun, K.; Liu, Y.; Chirvony, V.S.; Patra, A.; et al. Lead-Free Halide Perovskite Materials and Optoelectronic Devices: Progress and Prospective. *Adv. Funct. Mater.* **2023**. [CrossRef]
140. Toshniwal, A.; Kheraj, V. Development of Organic-Inorganic Tin Halide Perovskites: A Review. *Sol. Energy* **2017**, *149*, 54–59. [CrossRef]
141. Kim, D.Y.; Jung, J.G.; Lee, Y.J.; Park, M.H. Lead-Free Halide Perovskite Nanocrystals for Light-Emitting Diodes. *Materials* **2023**, *16*, 6317. [CrossRef] [PubMed]
142. Lai, M.L.; Tay, T.Y.S.; Sadhanala, A.; Dutton, S.E.; Li, G.; Friend, R.H.; Tan, Z.K. Tunable Near-Infrared Luminescence in Tin Halide Perovskite Devices. *J. Phys. Chem. Lett.* **2016**, *7*, 2653–2658. [CrossRef] [PubMed]
143. Yuan, F.; Zheng, X.; Johnston, A.; Wang, Y.; Zhou, C.; Dong, Y.; Chen, B.; Chen, H.; Fan, J.Z.; Sharma, G.; et al. Color-Pure Red Light-Emitting Diodes Based on Two-Dimensional Lead-Free Perovskites. *Sci. Adv.* **2020**, *6*, eabb0253. [CrossRef] [PubMed]
144. Dong, H.; Ran, C.; Gao, W.; Sun, N.; Liu, X.; Xia, Y.; Chen, Y.; Huang, W. Crystallization Dynamics of Sn-Based Perovskite Thin Films: Toward Efficient and Stable Photovoltaic Devices. *Adv. Energy Mater.* **2022**, *12*, 2102213. [CrossRef]
145. Meyer, E.; Mutukwa, D.; Zingwe, N.; Taziwa, R. Lead-Free Halide Double Perovskites: A Review of the Structural, Optical, and Stability Properties as Well as Their Viability to Replace Lead Halide Perovskites. *Metals* **2018**, *8*, 667. [CrossRef]
146. Gourji, F.H. Dhayalan Velauthapillai A Review on Cs-Based Pb-Free Double Halide Perovskites: From Theoretical and Experimental Studies to Doping and Applications. *Molecules* **2021**, *26*, 2010. [CrossRef] [PubMed]
147. Liu, Z.; Sun, Y.; Cai, T.; Yang, H.; Zhao, J.X.; Yin, T.; Hao, C.; Chen, M.; Shi, W.; Li, X.; et al. Two-Dimensional Cs2AgInxBi1-XCl6 Alloyed Double Perovskite Nanoplatelets for Solution-Processed Light-Emitting Diodes. *Adv. Mater.* **2023**, *35*, 2211235. [CrossRef]
148. Jiang, F.; Wu, Z.; Lu, M.; Gao, Y.; Li, X.; Bai, X.; Ji, Y.; Zhang, Y. Broadband Emission Origin in Metal Halide Perovskites: Are Self-Trapped Excitons or Ions? *Adv. Mater.* **2023**, *35*, 2211088. [CrossRef]
149. Arfin, H.; Kshirsagar, A.S.; Kaur, J.; Mondal, B.; Xia, Z.; Chakraborty, S.; Nag, A. Ns2Electron (Bi3+and Sb3+) Doping in Lead-Free Metal Halide Perovskite Derivatives. *Chem. Mater.* **2020**, *32*, 10255–10267. [CrossRef]
150. Li, S.; Luo, J.; Liu, J.; Tang, J. Self-Trapped Excitons in All-Inorganic Halide Perovskites: Fundamentals, Status, and Potential Applications. *J. Phys. Chem. Lett.* **2019**, *10*, 1999–2007. [CrossRef]
151. Luo, J.; Wang, X.; Li, S.; Liu, J.; Guo, Y.; Niu, G.; Yao, L.; Fu, Y.; Gao, L.; Dong, Q.; et al. Efficient and Stable Emission of Warm-White Light from Lead-Free Halide Double Perovskites. *Nature* **2018**, *563*, 541–545. [CrossRef] [PubMed]
152. Zhang, Y.; Zhang, Z.; Yu, W.; He, Y.; Chen, Z.; Xiao, L.; Shi, J.J.; Guo, X.; Wang, S.; Qu, B. Lead-Free Double Perovskite Cs2AgIn0.9Bi0.1Cl6 Quantum Dots for White Light-Emitting Diodes. *Adv. Sci.* **2022**, *9*, 2102895. [CrossRef] [PubMed]

153. Liu, Y.; Dong, Y.; Zhu, T.; Ma, D.; Proppe, A.; Chen, B.; Zheng, C.; Hou, Y.; Lee, S.; Sun, B.; et al. Bright and Stable Light-Emittin Diodes Based on Perovskite Quantum Dots in Perovskite Matrix. *J. Am. Chem. Soc.* **2021**, *143*, 15606–15615. [CrossRef] [PubMed
154. Kong, L.; Zhang, X.; Zhang, C.; Wang, L.; Wang, S.; Cao, F.; Zhao, D.; Rogach, A.L.; Yang, X. Stability of Perovskite Light-Emittin Diodes: Existing Issues and Mitigation Strategies Related to Both Material and Device Aspects. *Adv. Mater.* **2022**, *34*, 220521 [CrossRef] [PubMed]
155. Ye, Y.C.; Li, Y.Q.; Cai, X.Y.; Zhou, W.; Shen, Y.; Shen, K.C.; Wang, J.K.; Gao, X.; Zhidkov, I.S.; Tang, J.X. Minimizing Optical Energ Losses for Long-Lifetime Perovskite Light-Emitting Diodes. *Adv. Funct. Mater.* **2021**, *31*, 2105813. [CrossRef]
156. Xu, H.; Wang, X.; Li, Y.; Cai, L.; Tan, Y.; Zhang, G.; Wang, Y.; Li, R.; Liang, D.; Song, T.; et al. Prominent Heat Dissipatio in Perovskite Light-Emitting Diodes with Reduced Efficiency Droop for Silicon-Based Display. *J. Phys. Chem. Lett.* **2020**, *1* 3689–3698. [CrossRef] [PubMed]
157. Juarez-Perez, E.J.; Ono, L.K.; Qi, Y. Thermal Degradation of Formamidinium Based Lead Halide Perovskites into: Sym -Triazin and Hydrogen Cyanide Observed by Coupled Thermogravimetry-Mass Spectrometry Analysis. *J. Mater. Chem. A* **2019**, 16912–16919. [CrossRef]
158. Das, A.; Ghorai, A.; Saha, K.; Chatterjee, A.; Jeong, U. Various Approaches to Synthesize Water-Stable Halide PeNCs. *J. Mate Chem. A* **2023**, *11*, 6796–6813. [CrossRef]
159. Zhu, Z.; Hadjiev, V.G.; Rong, Y.; Guo, R.; Cao, B.; Tang, Z.; Qin, F.; Li, Y.; Wang, Y.; Hao, F.; et al. Interaction of Organic Catio with Water Molecule in Perovskite MAPbI3: From Dynamic Orientational Disorder to Hydrogen Bonding. *Chem. Mater.* **2016**, *2* 7385–7393. [CrossRef]
160. Leguy, A.M.A.; Hu, Y.; Campoy-Quiles, M.; Alonso, M.I.; Weber, O.J.; Azarhoosh, P.; Van Schilfgaarde, M.; Weller, M.T.; Bein, T Nelson, J.; et al. Reversible Hydration of CH3NH3PbI3 in Films, Single Crystals, and Solar Cells. *Chem. Mater.* **2015**, *27*, 3397–340 [CrossRef]
161. Zhang, H.; Fu, X.; Tang, Y.; Wang, H.; Zhang, C.; Yu, W.W.; Wang, X.; Zhang, Y.; Xiao, M. Phase Segregation Due to Ion Migratio in All-Inorganic Mixed-Halide Perovskite Nanocrystals. *Nat. Commun.* **2019**, *10*, 1088. [CrossRef] [PubMed]
162. Xiao, Z.; Yuan, Y.; Shao, Y.; Wang, Q.; Dong, Q.; Bi, C.; Sharma, P.; Gruverman, A.; Huang, J. Giant Switchable Photovoltaic Effec in Organometal Trihalide Perovskite Devices. *Nat. Mater.* **2015**, *14*, 193–197. [CrossRef] [PubMed]
163. Lee, S.; Park, J.H.; Lee, B.R.; Jung, E.D.; Yu, J.C.; Di Nuzzo, D.; Friend, R.H.; Song, M.H. Amine-Based Passivating Materials fo Enhanced Optical Properties and Performance of Organic-Inorganic Perovskites in Light-Emitting Diodes. *J. Phys. Chem. Let* **2017**, *8*, 1784–1792. [CrossRef] [PubMed]
164. Schulz, P.; Tiepelt, J.O.; Christians, J.A.; Levine, I.; Edri, E.; Sanehira, E.M.; Hodes, G.; Cahen, D.; Kahn, A. High-Work-Functior Molybdenum Oxide Hole Extraction Contacts in Hybrid Organic-Inorganic Perovskite Solar Cells. *ACS Appl. Mater. Interfaces* **2016**, *8*, 31491–31499. [CrossRef] [PubMed]
165. Wang, Y.; Teng, Y.; Lu, P.; Shen, X.; Jia, P.; Lu, M.; Shi, Z.; Dong, B.; Yu, W.W.; Zhang, Y. Low Roll-Off Perovskite Quantum Do Light-Emitting Diodes Achieved by Augmenting Hole Mobility. *Adv. Funct. Mater.* **2020**, *30*, 1910140. [CrossRef]
166. Kim, H.; Zhao, L.; Price, J.S.; Grede, A.J.; Roh, K.; Brigeman, A.N.; Lopez, M.; Rand, B.P.; Giebink, N.C. Hybrid Perovskite Ligh Emitting Diodes under Intense Electrical Excitation. *Nat. Commun.* **2018**, *9*, 4893. [CrossRef] [PubMed]
167. Fakharuddin, A.; Qiu, W.; Croes, G.; Devižis, A.; Gegevičius, R.; Vakhnin, A.; Rolin, C.; Genoe, J.; Gehlhaar, R.; Kadashchuk A.; et al. Reduced Efficiency Roll-Off and Improved Stability of Mixed 2D/3D Perovskite Light Emitting Diodes by Balancing Charge Injection. *Adv. Funct. Mater.* **2019**, *29*, 1904101. [CrossRef]
168. Zou, W.; Li, R.; Zhang, S.; Liu, Y.; Wang, N.; Cao, Y.; Miao, Y.; Xu, M.; Guo, Q.; Di, D.; et al. Minimising Efficiency Roll-off in High-Brightness Perovskite Light-Emitting Diodes. *Nat. Commun.* **2018**, *9*, 608. [CrossRef]
169. Klimov, V.I.; Mikhailovsky, A.A.; McBranch, D.W.; Leatherdale, C.A.; Bawendi, M.G. Quantization of Multiparticle Auger Rates in Semiconductor Quantum Dots. *Science* **2000**, *287*, 1011–1014. [CrossRef]
170. Bi, C.; Yao, Z.; Hu, J.; Wang, X.; Zhang, M.; Tian, S.; Liu, A.; Lu, Y.; de Leeuw, N.H.; Sui, M.; et al. Suppressing Auger Recombination of Perovskite Quantum Dots for Efficient Pure-Blue-Light-Emitting Diodes. *ACS Energy Lett.* **2023**, *8*, 731–739 [CrossRef]
171. Chen, F.; Liu, Y.; Zhang, D.; Jiang, X.; Cai, P.; Si, J.; Hu, Q.; Fang, Z.; Dai, X.; Song, J.; et al. Bilayer Phosphine Oxide Modification toward Efficient and Large-Area Pure-Blue Perovskite Quantum Dot Light-Emitting Diodes. *Sci. Bull.* **2023**, *68*, 2354–2361 [CrossRef] [PubMed]
172. Tang, Y.Y.; Shen, Y.; Yu, Y.; Zhang, K.; Wang, B.F.; Tang, J.X.; Li, Y.Q. Comprehensive Crystal Regulation Reduces Interfacial Energy Loss for Efficient Blue Perovskite Light-Emitting Diodes. *Small* **2023**, 2309309. [CrossRef] [PubMed]
173. Zhou, Y.H.; Wang, C.; Yuan, S.; Zou, C.; Su, Z.; Wang, K.L.; Xia, Y.; Wang, B.; Di, D.; Wang, Z.K.; et al. Stabilized Low-Dimensional Species for Deep-Blue Perovskite Light-Emitting Diodes with EQE Approaching 3.4%. *J. Am. Chem. Soc.* **2022**, *144*, 18470–18478 [CrossRef] [PubMed]
174. Boehme, S.C.; Bodnarchuk, M.I.; Burian, M.; Bertolotti, F.; Cherniukh, I.; Bernasconi, C.; Zhu, C.; Erni, R.; Amenitsch, H.; Naumenko, D.; et al. Strongly Confined CsPbBr3 Quantum Dots as Quantum Emitters and Building Blocks for Rhombic Superlattices. *ACS Nano* **2023**, *17*, 2089–2100. [CrossRef] [PubMed]
175. Wang, Y.; Jia, F.; Li, X.; Teale, S.; Xia, P.; Liu, Y.; Chan, P.T.; Wan, H.; Hassan, Y.; Imran, M.; et al. Self-Assembled Monolayer—Based Blue Perovskite LEDs. *Sci. Adv.* **2023**, *9*, eadh2140. [CrossRef]

176. Yuan, S.; Cui, L.S.; Dai, L.; Liu, Y.; Liu, Q.W.; Sun, Y.Q.; Auras, F.; Anaya, M.; Zheng, X.; Ruggeri, E.; et al. Efficient and Spectrally Stable Blue Perovskite Light-Emitting Diodes Employing a Cationic π-Conjugated Polymer. *Adv. Mater.* **2021**, *33*, 2103640. [CrossRef]
177. Feng, W.; Zhao, Y.; Lin, K.; Lu, J.; Liang, Y.; Liu, K.; Xie, L.; Tian, C.; Lyu, T.; Wei, Z. Polymer-Assisted Crystal Growth Regulation and Defect Passivation for Efficient Perovskite Light-Emitting Diodes. *Adv. Funct. Mater.* **2022**, *32*, 2203371. [CrossRef]
178. Li, M.; Sun, R.; Chang, J.; Dong, J.; Tian, Q.; Wang, H.; Li, Z.; Yang, P.; Shi, H.; Yang, C.; et al. Orientated Crystallization of FA-Based Perovskite via Hydrogen-Bonded Polymer Network for Efficient and Stable Solar Cells. *Nat. Commun.* **2023**, *14*, 573. [CrossRef]
179. Huang, Z.; Li, L.; Wu, T.; Xue, T.; Sun, W.; Pan, Q.; Wang, H.; Xie, H.; Chi, J.; Han, T.; et al. Wearable Perovskite Solar Cells by Aligned Liquid Crystal Elastomers. *Nat. Commun.* **2023**, *14*, 1204. [CrossRef]
180. Shen, Y.; Xu, G.; Li, J.; Lin, X.; Yang, F.; Yang, H.; Chen, W.; Wu, Y.; Wu, X.; Cheng, Q.; et al. Functional Ionic Liquid Polymer Stabilizer for High-Performance Perovskite Photovoltaics. *Angew. Chem. Int. Ed.* **2023**, *62*, e202300690. [CrossRef]
181. Yang, J.N.; Song, Y.; Yao, J.S.; Wang, K.H.; Wang, J.J.; Zhu, B.S.; Yao, M.M.; Rahman, S.U.; Lan, Y.F.; Fan, F.J.; et al. Potassium Bromide Surface Passivation on CsPbI3-XBrx Nanocrystals for Efficient and Stable Pure Red Perovskite Light-Emitting Diodes. *J. Am. Chem. Soc.* **2020**, *142*, 2956–2967. [CrossRef] [PubMed]
182. Zhang, J.; Cai, B.; Zhou, X.; Yuan, F.; Yin, C.; Wang, H.; Chen, H.; Ji, X.; Liang, X.; Shen, C.; et al. Ligand-Induced Cation–π Interactions Enable High-Efficiency, Bright, and Spectrally Stable Rec. 2020 Pure-Red Perovskite Light-Emitting Diodes. *Adv. Mater.* **2023**, *35*, 2303938. [CrossRef] [PubMed]
183. Ren, Z.; Xiao, X.; Ma, R.; Lin, H.; Wang, K.; Sun, X.W.; Choy, W.C.H. Hole Transport Bilayer Structure for Quasi-2D Perovskite Based Blue Light-Emitting Diodes with High Brightness and Good Spectral Stability. *Adv. Funct. Mater.* **2019**, *29*, 1905339. [CrossRef]
184. Quan, L.N.; García de Arquer, F.P.; Sabatini, R.P.; Sargent, E.H. Perovskites for Light Emission. *Adv. Mater.* **2018**, *30*, 1801996. [CrossRef]
185. Yang, X.; Ma, L.; Yu, M.; Chen, H.; Ji, Y.; Hu, A.; Zhong, Q.; Jia, X.; Wang, Y.; Zhang, Y.; et al. Focus on Perovskite Emitters in Blue Light-Emitting Diodes. *Light Sci. Appl.* **2023**, *12*, 177. [CrossRef]

coatings

MDPI

Article

Modeling Enthalpy of Formation with Machine Learning for Structural Evaluation and Thermodynamic Stability of Organic Semiconductors

Sadaf Noreen [1], Mamduh J. Aljaafreh [2,*] and Sajjad H. Sumrra [1]

[1] Department of Chemistry, University of Gujrat, Gujrat 50700, Punjab, Pakistan; sadafnoreen234@gmail.com (S.N.); sajjadchemist@uog.edu.pk (S.H.S.)

[2] Physics Department, College of Science, Imam Mohammad Ibn Saud Islamic University (IMSIU), Riyadh 11623, Saudi Arabia

* Correspondence: maljaafreh@imamu.edu.sa

Abstract

The enthalpy of formation (dH_m) is a crucial parameter in evaluating the structural stability of organic semiconductors. In this study, we employed machine learning (ML) models to predict the dH_m of organic semiconductors. The current results show that Kappa2 and NumRotableBonds are highly correlated with the dH_m, indicating their importance in determining the stability of these materials. Using Gradient Boosting, Random Forest, and Extra Trees models, we achieved a high R^2 value of 0.68–0.70, demonstrating the effectiveness of these models in predicting the dH_m. Further analysis using SHAP values revealed that Kappa2 and fr_uncrch_alkane are the most important descriptors in determining the dH_m. These findings provide valuable insights into the structural evaluation and stability of organic semiconductors and highlight the potential of ML models in predicting key properties of these materials.

Keywords: enthalpy of formation; machine learning; SHAP values; random forest; Kappa2

1. Introduction

Organic semiconductors have gained significant attention in recent years for their potential applications in light harvesting, owing to their unique combination of properties [1]. Their importance in this context lies in their ability to efficiently absorb light and convert it into electrical energy, thereby facilitating the development of highly efficient solar cells and photovoltaic devices [2]. They also exhibit superior flexibility, low cost, and facile processability [3], which enables the fabrication of large area, flexible, and lightweight photovoltaic devices that can be easily integrated into a variety of applications [4]. The molecular structure of organic semiconductors can be tailored to optimize their optical and electrical properties [5], allowing for the design of materials with specific absorption spectra and charge transport characteristics [6]. Additionally, the use of organic semiconductors in light harvesting applications offers the potential for the development of novel device architectures, such as tandem solar cells and photodetectors [7], which can further enhance the efficiency and versatility of these devices [8]. As a result, organic semiconductors are poised to play a pivotal role in the development of next-generation light harvesting technologies [9], enabling the creation of innovative, high-performance devices that can efficiently capture and convert sunlight into usable energy [10].

The thermal properties of organic semiconductors are of paramount importance in determining their suitability for various applications, particularly in the design of thin films and devices that operate under diverse thermal conditions [11]. One crucial aspect of their thermal properties is the dH_m, which is a measure of the energy released or absorbed during the formation of a material from its constituent elements [12]. Organic semiconductors typically exhibit relatively low enthalpies of formation, which facilitates their deposition and crystallization into thin films with minimal thermal damage or degradation [13]. This property enables the design and fabrication of thin films with precise control over their morphology, crystal structure, and intermolecular interactions, which is essential for optimizing their electronic and optoelectronic properties [14]. In addition, the stability-driven features of materials, which may be characterized by their glass transition temperature, melting point, and thermal decomposition temperature, are critical in determining their reliability and durability under various operating conditions [15]. By carefully tailoring the molecular structure and processing conditions, it is possible to design organic semiconductors with optimized thermal properties [16], allowing for the fabrication of thin films and devices that can withstand elevated temperatures, thermal stress, and other environmental factors [17]. The use of high-enthalpy of formation materials can also enable the creation of thin films with enhanced thermal stability, while low-enthalpy of formation materials may be more suitable for applications requiring low-temperature processing or flexible substrates.

The integration of ML techniques has revolutionized the field of materials science, particularly in predicting relevant properties of organic semiconductors [18]. The ability to accurately predict properties such as dH_m, thermal stability [19], and electronic conductivity is crucial for the design and development of high-performance devices [20]. Its different algorithms can be trained on large datasets of experimental and computational results to identify complex relationships between the molecular structure and material properties, enabling the prediction of these properties with high accuracy [21]. This is particularly significant in the context of organic semiconductors, where the intricate relationships between molecular structure, crystal packing, and material properties make experimental characterization and traditional computational modeling approaches challenging [22]. The aim of this work is to utilize ML techniques to predict the dH_m of organic semiconductors, with a specific focus on understanding the structural basis of this property. By analyzing the relationships between molecular features, such as functional groups, molecular topology, and crystal structure, and the corresponding dH_m values, it is possible to gain insights into the underlying factors that influence this property. The predicted dH_m values can then be used to design and optimize organic semiconductors with tailored thermal properties, which is essential for the development of high-performance devices that can operate efficiently and reliably under various conditions. The structural basis of the dH_m can provide valuable information on the molecular design principles and processing conditions required to achieve optimal material properties, ultimately facilitating the discovery of new organic semiconductors with improved performance and stability. By leveraging ML techniques, this work aims to establish a robust framework for the prediction and understanding of dH_m in organic semiconductors, which can be extended to other material properties and applications, thereby accelerating the development of novel materials and devices (Figure 1). Our work presents a significant contribution to the field of organic electronics by introducing a machine learning-based approach for predicting the dH_m and evaluating the structural and thermodynamic stability of organic semiconductors, enabling the accelerated discovery of stable and efficient materials.

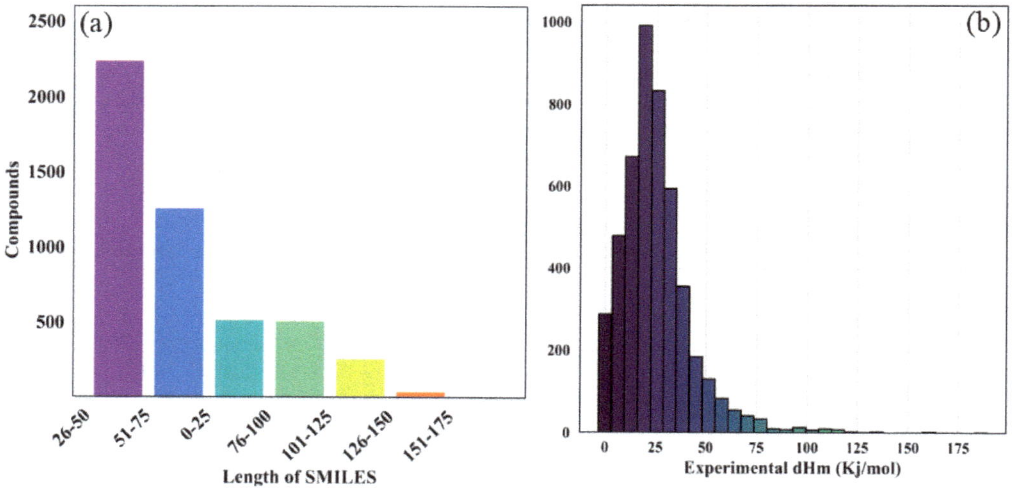

Figure 1. A view of the (**a**) length of SMILES and (**b**) experimental dH_m of organic semiconductor data.

2. Methodology

2.1. ML Analysis

For relevant ML calculations, the latest Python program (V 3.10.14) and its latest libraries were imported. For this, all the dataset files were imported by their Pandas module [23], while the NumPy [24] and RDKit [25] libraries were imported for their descriptor designing. Their calculated results were visualized by its Matplotlib library (V3.10.3), while their scientific/statistical calculations were performed by the Scikit-learn module [26]. Their relevant plots and graphs were plotted by importing the Matplotlib library. Their molecular descriptors from their experimental results with their SMILES length (Figure 2) were designed by importing the RDKit toolkit [27]. Their descriptors were designed to realize their impact on their target property (d_{Hm}). An example of a few prominent descriptor relations is given below as Equations (1)–(4):

$$MW = \sum_{i=1}^{n} m_i \quad (1)$$

$$LogP = Log_{10}\left(\frac{C_{octanol}}{C_{water}}\right) \quad (2)$$

$$W = \sum_{i=1}^{n} \sum_{j=1}^{n} d_{ij} \quad (3)$$

$$EN = \frac{1}{n}\sum_{i=1}^{n} x_i \quad (4)$$

To evaluate their different molecular connectivity descriptors (χ_n^v), their hydrogen-suppressed skeletons were employed. Their valence connectivity index (χ_o^v) [28] was designed from their non-hydrogen atoms having distinct δ^v values, as denoted by their atomic valence delta (δ^v) [29] (Equation (5)).

$$\chi_o^v = \sum_{i=10}^{n} (\delta^v)^{-0.6} \quad (5)$$

Each of their non-hydrogen atoms with their atomic number (Z) and valence electrons (Z^V), along with their attached hydrogen atoms (h), calculated their δ^V after combining them (Equation (6)).

$$\delta = \frac{Z^h - h}{Z - Z^h - 1} \tag{6}$$

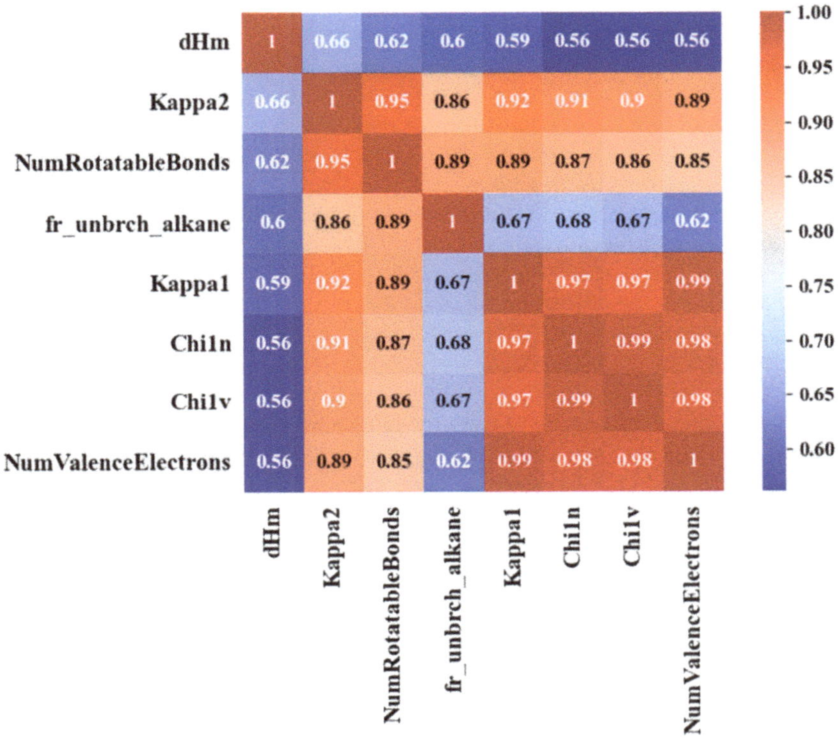

Figure 2. A view of the Pearson correlation heatmap of the top correlating descriptors with the target property (d_{Hm}).

2.2. Correlations and Feature Scores

The Pearson correlation of the target property with its designed descriptors as their coefficient (r) [30] was quantified through their linear relationship by taking their two variables, such as XX and YY (Equation (7)).

$$r = \frac{n(\sum XY) - (\sum X)(\sum Y)}{\sqrt{\left[n\sum X^2 - (\sum X)^2\right]}\sqrt{\left[n\sum Y^2 - (\sum Y)^2\right]}} \tag{7}$$

while n denoted their data point numbers, where $\sum XY$ was the sum of the product of their paired scores (Equation (8)). Also, their feature importance scores (FI_j) for a feature j were also calculated by their evaluated models [31].

$$FI_j = \frac{1}{T}\sum_{t=1}^{T} \Delta I_{jt} \tag{8}$$

where T is the total number of trees in the model, while ΔI_{jt} shows their decrease in their impurity of a tree (t). A tree-based model can be evaluated for its decrease in impurity, like Gini impurity or entropy [32], by bringing each feature of t.

3. Results and Discussion

3.1. Descriptor Designing

The dH_m, a measure of the energy change associated with the formation of a compound from its constituent elements, has been found to have significant correlations with several molecular descriptors. The highest correlation is with kappa2 [33], a topological index that characterizes the shape of a molecule, with a correlation coefficient of 0.66 (Figure 2). This suggests that the shape of a molecule plays a crucial role in determining its dH_m. The polymers with more complex shapes, as indicated by higher kappa2 values, tend to have higher enthalpies of formation, possibly due to the increased energy required to form and stabilize these structures. Another molecular descriptor that showed a strong correlation with dH_m is NumRotatbleBonds, which represents the number of rotatable bonds in a molecule, with a correlation coefficient of 0.62. This indicated that molecules with more rotatable bonds tend to have higher enthalpies of formation.

This could be rationalized by considering the fact that rotatable bonds introduce greater flexibility and conformational complexity to a molecule, which can lead to increased energy requirements for its formation. Furthermore, the correlation with fr_unbrch_alkane, which represents the fraction of unbranched alkane-like structures in a molecule, with a coefficient of 0.60 suggested that molecules with more linear, unbranched structures tend to have lower enthalpies of formation. The correlation with kappa1, another topological index, with a coefficient of 0.59 further reinforces the idea that molecular shape and topology play important roles in determining the dH_m. Additionally, the correlations with Chi1n and Chi1v, which were valence and non-valence connectivity indices, respectively, with coefficients around 0.56 suggested that the distribution of electrons and the overall connectivity of a molecule also influence its dH_m. Similarly, the correlation with NumValenceElectrons, which represents the total number of valence electrons in a molecule, with a coefficient of 0.56 indicated that the electron density and valence electron configuration of a molecule contribute to its dH_m. All these correlations suggest that the dH_m is a complex property that is influenced by a combination of molecular shape, topology, flexibility, and electronic structure.

3.2. Model Evaluation

The evaluation of ensemble and non-ensemble ML models to estimate their performance against experimental d_{Hm} values is a comprehensive process that involves assessing the predictive capabilities of various algorithms. Among the models evaluated, Gradient Boosting (GB) [34], Random Forest (RF) [35], Extra Trees [36], and Historical Gradient Boosting (Hist GB) [37] were notable for their ability to accurately predict d_{Hm} values. The GB is a powerful ensemble learning regressor that combines multiple weak models to create a strong predictive model, working by iteratively training decision trees, with each subsequent tree attempting to correct the errors of the previous one (Figure 3). RF and Extra Trees are also ensemble methods that are based on the concept of decision trees, with RF combining multiple trees to improve accuracy and robustness and Extra Trees using a technique called "averaging" to combine the predictions of individual trees. The Hist GB, on the other hand, incorporates temporal information into the model, allowing it to capture patterns and trends in the data over time. The evaluation of these models against experimental d_{Hm} values involves comparing their predicted values to the actual values obtained from experiments, using metrics such as mean absolute error, mean squared error, and coefficient of determination.

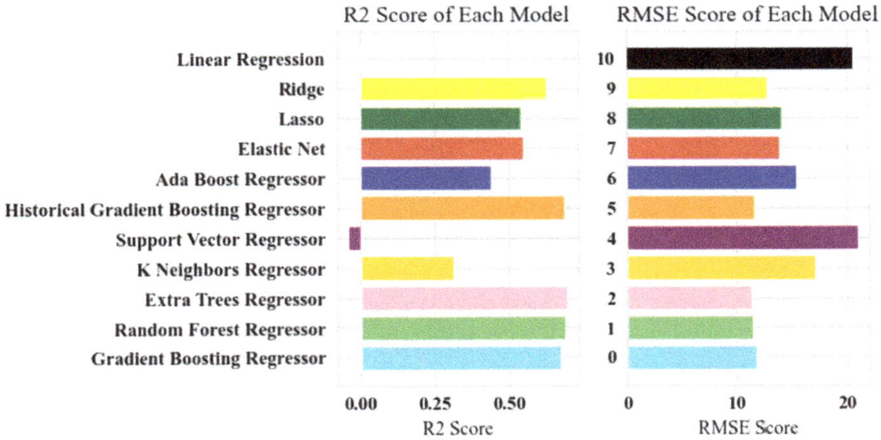

Figure 3. A view of the results of model evaluation by taking the dH_m as target properties.

The performance of the four best regression models, namely GB, RF, Extra Trees, and Hist GB, can be evaluated based on their R^2 and RMSE values. The R^2 value, also known as the coefficient of determination, measures the proportion of the variance in the dependent variable that is predictable from the independent variable(s). The RMSE value, on the other hand, measures the average magnitude of the errors in the model's predictions. Among the four models, RF has the highest R^2 value of 0.70, indicating that it is able to explain 70% of the variance in the dependent variable. This suggested that RF was the most effective model in terms of capturing the underlying patterns and relationships in the data.

In terms of RMSE, RF also has the lowest value of 11.27, indicating that its predictions were, on average, closest to the actual values. This could be a significant finding, as it suggested that RF was not only able to capture the underlying patterns in the data but could also be able to make accurate predictions (Figure 4). The Extra Trees and Hist GB regressors had their identical R^2 and RMSE values of 0.69 and 11.50, respectively, suggesting that they could be equally effective in terms of explaining the variance in the dependent variable and making accurate predictions. GB, on the other hand, has the lowest R^2 value of 0.68 and the highest RMSE value of 11.68, indicating that it could be the least effective model among the four.

The performance of these models can be attributed to their respective strengths and weaknesses. The RF regressor, for example, is known for its ability to handle high-dimensional data and to reduce overfitting, which might explain its strong performance in this case. The Extra Trees and Hist GB regressors, on the other hand, are both ensemble methods that combine multiple models to improve their predictive performance, which might explain their similar performance. The GB regressor, while a powerful algorithm, might be more prone to overfitting, which may explain its relatively poor performance.

The implications of these findings could be significant in suggesting that RF might be the most suitable model for predicting d_{Hm} values in this particular context [38]. This is because RF is able to capture the underlying patterns and relationships in the data and make accurate predictions, which is critical in many applications. Furthermore, the fact that Extra Trees and Hist GB have similar performance to RF suggests that they might also be suitable alternatives, particularly if there are specific requirements or constraints that need to be considered [39]. In contrast, the GB regressor might require further tuning or modification to improve its performance, which might involve adjusting its hyperparameters or using techniques such as regularization to reduce overfitting. The results of this analysis could have important implications for the development of predictive models for d_{Hm} values and highlight the importance of carefully evaluating and comparing the performance of different models.

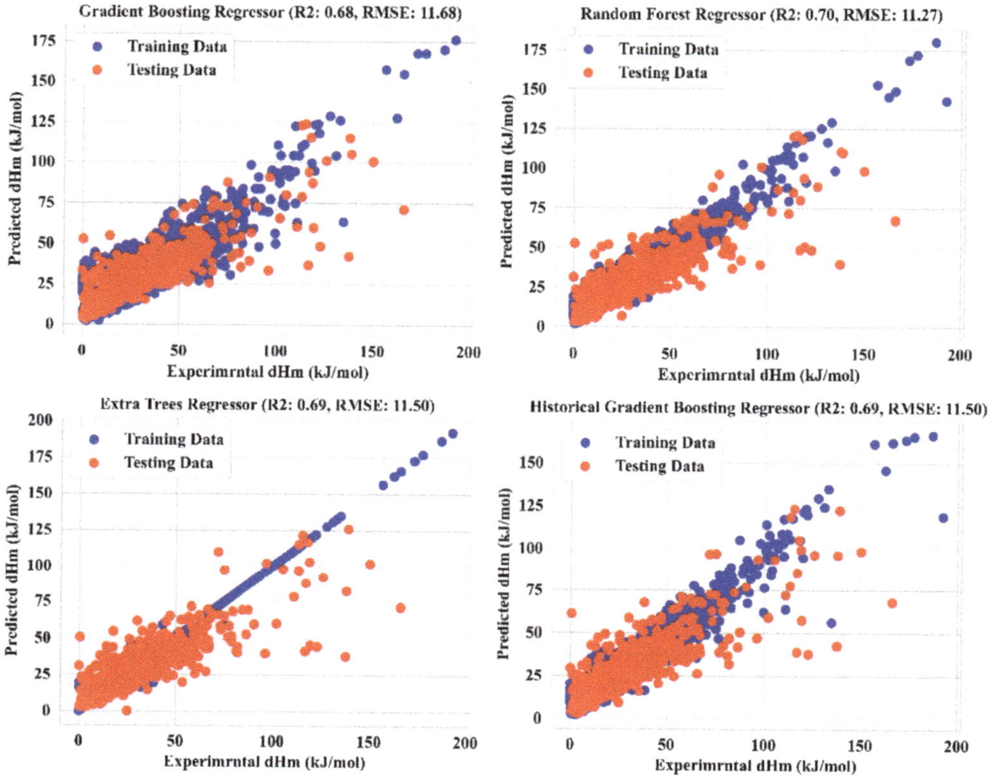

Figure 4. Scatter plots of experimental versus predicted d_{Hm} values against the top 4 best-performing models.

The density of residuals for each of the four models provides valuable insights into their performance and behavior [40]. The residuals, which represented the differences between the actual and predicted d_{Hm} values, were spread out across a range of values, indicating that each model could be capable of capturing a significant portion of the variability in the data. However, the specific characteristics of the residual distributions vary across the models. For example, GB and RF had similar residual ranges, spanning from −40 to 100, and predicted d_{Hm} values ranging from 0 to 175 (Figure 5). This suggested that those two models could be capable of handling a similar range of values and that their predictions could be centered around a similar range. In contrast, Extra Trees and Hist GB had slightly different residual characteristics. Extra Trees had a broader residual range, spanning from −60 to 100, and predicted d_{Hm} values ranging from 0 to 200. That indicated that Extra Trees was capable of handling a wider range of values and that its predictions were more spread out. The Hist GB, on the other hand, had a similar residual range to Extra Trees, spanning from −60 to 100, but its predicted d_{Hm} values are more limited, ranging from 0 to 150. This suggested that Hist GB was capable of handling a similar range of values to Extra Trees, but its predictions were more conservative.

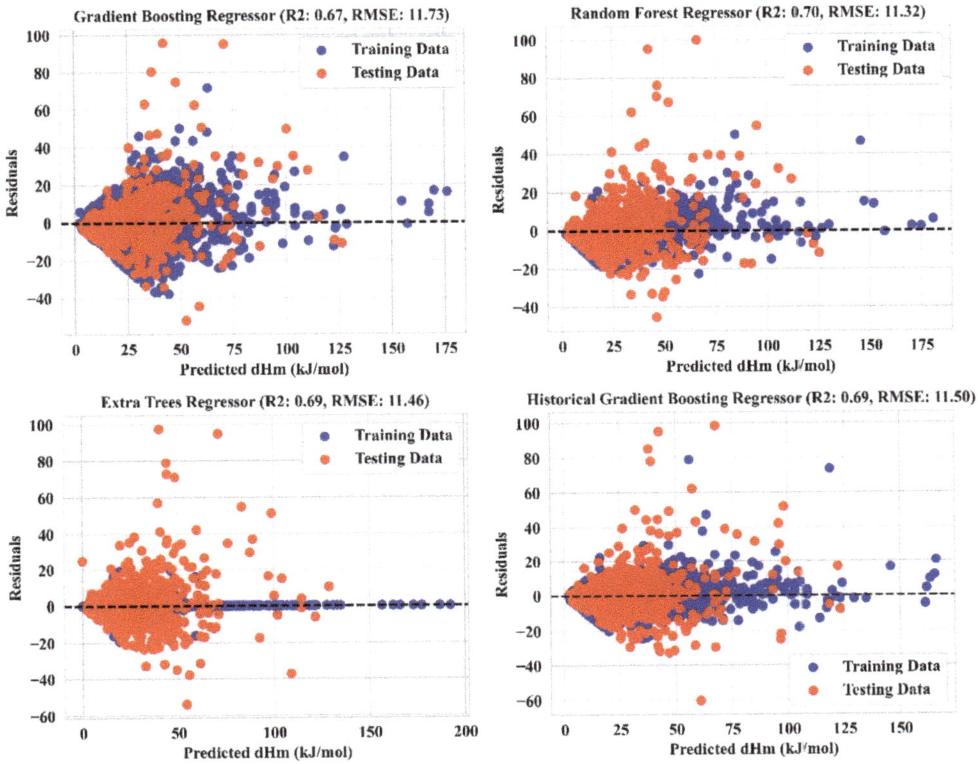

Figure 5. A view of the plots of density of residuals against their predicted d_{Hm} values against the top 4 best-performing models.

3.3. SHapley Impact

The SHAP value beeswarm plot of the evaluated model reveals valuable insights into the relative importance of each descriptor in predicting the d_{Hm} values. According to the plot, kappa2 emerged as the descriptor with the highest impact, indicating that it was the most influential feature in determining its model predictions. This is not surprising, given that kappa2 is a molecular descriptor that captures the spatial arrangement of atoms within a molecule, which can have a significant impact on its thermodynamic properties [41], including the heat of melting (Figure 6). The fact that kappa2 is the most important descriptor suggested that the model could be heavily reliant on the spatial arrangement of atoms in the molecule to make predictions. The next most important descriptor, fr_unbrch_alkane, was a functional group count descriptor that represents the number of unbranched alkane groups in the molecule. The presence of this descriptor in the top two suggested that the model was also considering the molecular structure and functional groups present in the molecule when making predictions. The fact that fr_unbrch_alkane followed kappa2 in importance indicated that the model was using a combination of spatial arrangement and functional group information to predict the d_{Hm} values. The third most important descriptor, kappa1, was another molecular descriptor that captures the spatial arrangement of atoms but with a slightly different focus than kappa2. The presence of kappa1 in the top three suggested that the model was using multiple molecular descriptors to capture different aspects of the molecular structure.

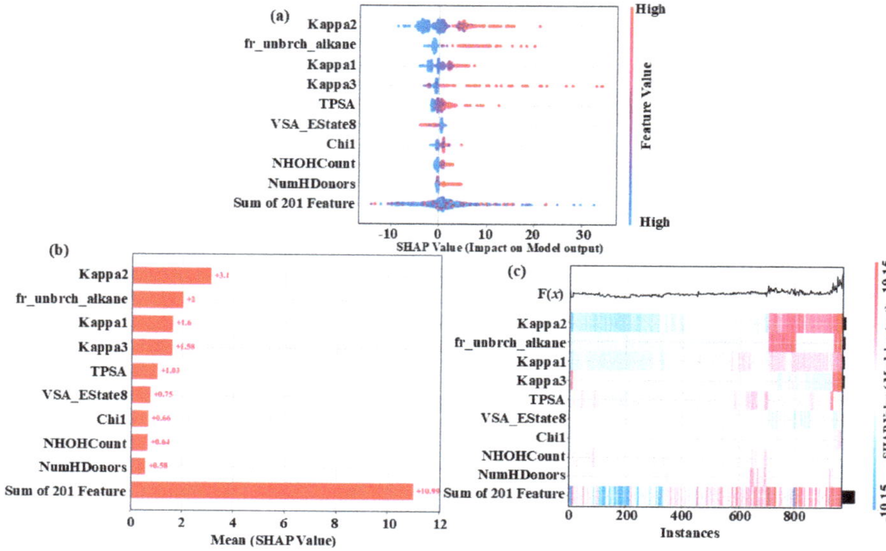

Figure 6. A view of the (**a**) SHAP values beeswarm plot, its (**b**) feature importance, and (**c**) descriptor impact instances of the dataset during the best evaluated model.

The remaining descriptors in the top list, including kappa3, TPSA, VSA_Estate8, and Chi1, also provided valuable insights into the model behavior. Kappa3 was another molecular descriptor that captured the spatial arrangement of atoms but with a focus on their molecular shape and size. TPSA, or topological polar surface area, was a descriptor that represented the surface area of the molecule that was accessible to a solvent. VSA_Estate8 was a descriptor that represented the van der Waals surface area of the molecule, while Chi1 was a descriptor that captured the molecular connectivity. The presence of these descriptors in the top list suggested that the model was using a combination of molecular structure, functional group, and surface area information to predict d_{Hm} values. Overall, the SHAP values beeswarm plot provided a detailed understanding of the model behavior and the relative importance of each descriptor in predicting d_{Hm} values.

3.4. Cross Validation

The K-fold cross-validation analysis for the model reveals a comprehensive evaluation of its performance across different folds [42]. The analysis was conducted for zero to four folds (five folds), where the data were divided into five folds and the model was trained and tested on each fold separately. Their training R^2 values ranged from 0.72 to 0.77, indicating that the model was able to capture the underlying patterns in the data and fit the training data well. The test R^2 values, which were a more reliable indicator of the model performance, ranged from 0.55 to 0.67. While the test R^2 values were lower than the training R^2 values, they were still relatively high, indicating that the model could be able to generalize well to unseen data (Figure 7). The fact that the test R^2 values were consistent across the different folds suggested that the model was robust and not sensitive to the specific data used for training and testing. The highest test R^2 value of 0.67 was observed in fold four, which suggested that the model might be overfitting to the training data in this fold. However, the difference between the training and test R^2 values was not significant enough to indicate severe overfitting. The K-fold cross-validation analysis also provided insight into the model performance across different folds. Fold one had the lowest test R^2 value of 0.55, which might indicate that the model was struggling to generalize to this particular set of data (Figure 7). In contrast, fold four had the highest test R^2 value of 0.67, which suggested that the model was performing well on this set of data. The fact that the model performance varied across the different folds might indicate that the data were not perfectly representative of the underlying population or that the model was sensitive to certain characteristics of the data. The K-fold cross-validation analysis suggested that the model was performing well, with high R^2 values for both training and testing sets. The consistency of the test R^2 values across the different folds indicated that the model could be robust and able to generalize well to unseen data. However, the variation in the model performance across the different folds might indicate that further refinement of the model could be necessary to improve its performance and robustness. Additionally, the analysis highlighted the importance of using K-fold cross-validation to evaluate the model performance, as it provided a more comprehensive understanding of the model strengths and weaknesses than a single train–test split.

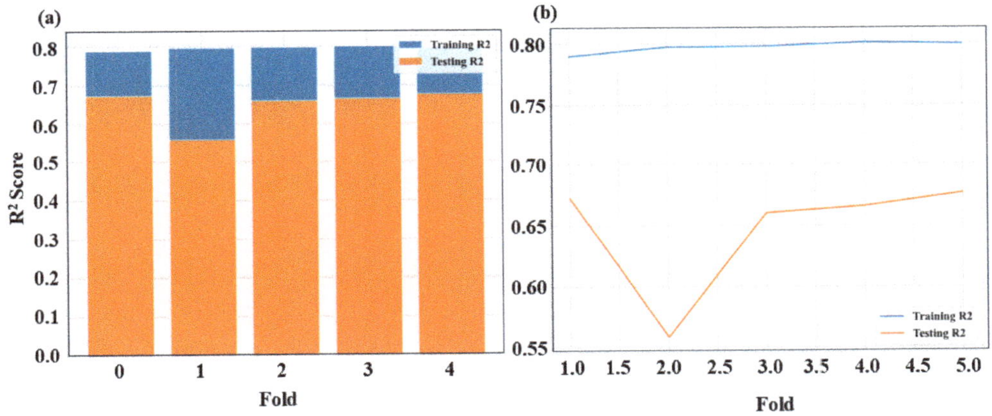

Figure 7. A view of the (**a**) bar graph showing the distribution of their R^2 for different folds, and a comparative view of their (**b**) R^2 for the training and testing datasets.

3.5. Hyperparameter Tuning

The hyperparameter tuning of the model revealed a comprehensive analysis of the learning rate impact on its performance [43]. The results showed that the model test score was highly sensitive to the learning rate, with a significant range of values resulting in varying degrees of performance. For learning rates between −25 and 1.25, the test score ranged up to 0.7, indicating that the model was capable of achieving high accuracy when the learning rate was optimized (Figure 8).

The most notable finding was that a learning rate of up to 0.50 resulted in a test score of 0.7, which suggested that the model was able to learn effectively and generalize well to unseen data when the learning rate was moderate. This is a significant result, as it indicates that the model is robust and can achieve high performance even when the learning rate is not extremely high (Figure 8). The fact that the test score remains high for learning rates up to 0.50 suggested that the model could be able to adapt to a range of learning rates and find the optimal solution. However, for learning rates 0.05–0.80, the mean test score ranged from 0.62, indicating that the model performance was not as consistent across that range. This also suggested that the model might be overfitting or underfitting for certain learning rates, resulting in suboptimal performance.

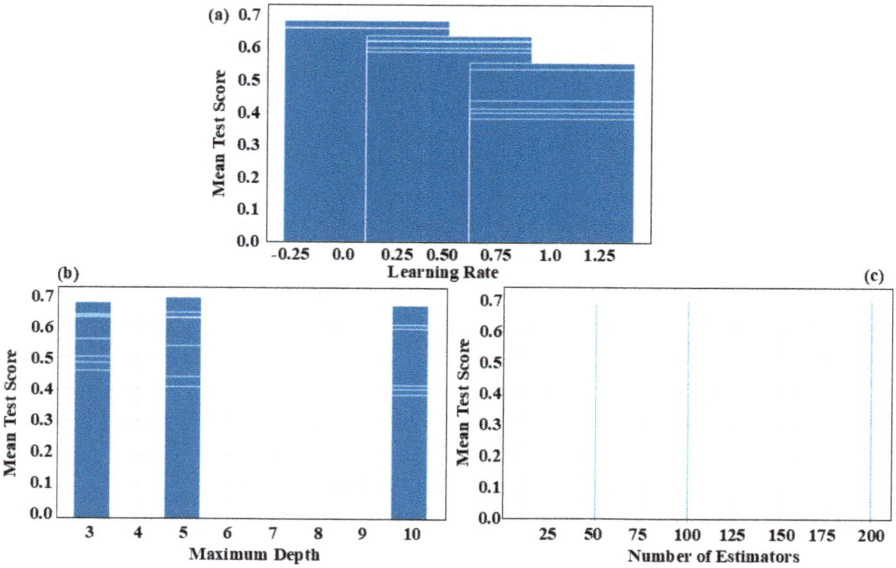

Figure 8. A view of the different hyperparameter tuning results of the models, including their (**a**) learning rate, (**b**) maximum depth, and (**c**) number of estimators.

The fact that the mean test score was lower than the maximum test score achieved at a learning rate of 0.50 suggested that the model performance was highly dependent on the learning rate and that finding the optimal learning rate was important to achieve their accuracy. For learning rates of 0.55–1.25, the test score dropped to 0.53, indicating that the model performance was negatively impacted by higher learning rates. This suggested that the model could be sensitive to overfitting when the learning rate was too high, resulting in poor generalization to unseen data. The fact that the test score dropped off significantly for learning rates above 0.55 suggested that the model performance was highly dependent on the learning rate and that finding the optimal learning rate was crucial for achieving high accuracy. The maximum depth of the model varied up to 10, with notable test scores at different depths. The maximum test score of 0.62 was achieved at a depth of 3, indicating that the model was able to capture relevant features at this depth. A higher test score of 0.67 was achieved at a depth of 5, suggesting that the model was able to learn more complex patterns at this depth. However, the mean test score at a depth of 10 was 0.65, which was slightly lower than the maximum test score at a depth of 5, indicating that increasing the depth beyond 5 may not necessarily lead to better performance.

3.6. Data Clustering

The data clustering pattern of the data, in the form of its T-distributed Stochastic Neighbor Embedding (t-SNE) [44], revealed a complex and interesting structure, with two components ranging from −50 to 50 and −75 to 75, respectively. This showed that the data were not randomly distributed but rather formed clusters or groups that were separated by certain characteristics. Component 1 appeared to be the primary driver of the clustering pattern (Figure 9). The data points that fell within that range were densely packed, forming a large cluster in the center of the plot. This cluster was flanked by two smaller clusters, one on either side, which were separated from the central cluster by a gap/void in the data. The presence of those smaller clusters suggested that there might be sub-populations within the data that were distinct from the main cluster. Component 2 appeared to be secondary to component 1 in terms of its influence on the clustering pattern. However, it was still an important factor, as it could help to separate the data points into distinct groups or clusters. The data points that fell within the lower range of component 2 appeared to form a separate cluster that was distinct from the main cluster.

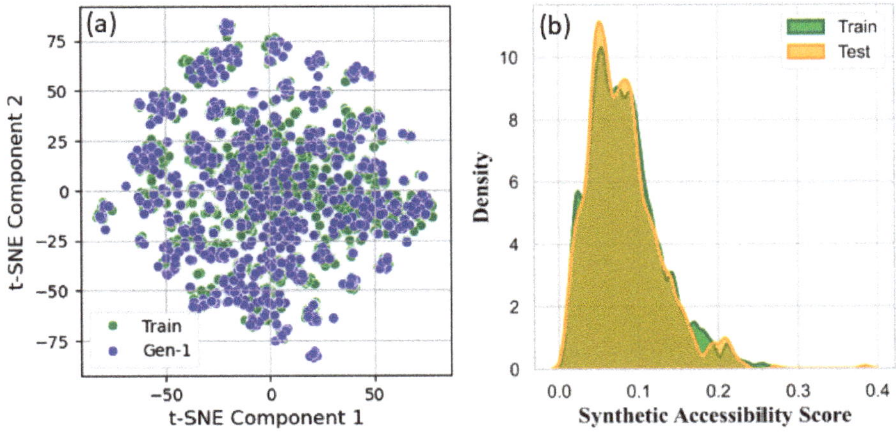

Figure 9. A view of the (**a**) data clustering analysis in the form of t-SNE maps, and (**b**) their synthetic accessibility score graphs for the training and testing datasets.

Similarly, the data points that fell within the upper range of component 2 also appeared to form a separate cluster. The combination of these two components resulted in a clustering pattern that was complex and multifaceted. The data points were not randomly distributed but rather formed a series of clusters and sub-clusters that were separated by gaps/voids in the data. That suggested that the data could not simply be a collection of random points but rather a structured dataset that contained hidden patterns and relationships. One possible interpretation of this clustering pattern could be that it reflected the presence of different sub-populations or groups within the data. For example, in a dataset of materials, the clusters might represent different types of materials, such as metals, polymers, or ceramics. The gaps/voids in the data might represent the boundaries between these different groups or the lack of data points in certain regions of the plot. Another possible interpretation could be that the clustering pattern reflected the presence of different characteristics or features within the data. For example, in a dataset of materials, the clusters might represent different properties or characteristics, such as strength, conductivity, or optical properties. The gaps or voids in the data might represent the absence of certain characteristics or features in certain regions of the plot.

4. Conclusions

This study demonstrated the potential of ML models in predicting the dH_m of organic semiconductors. The results showed that Kappa2 and NumRotableBonds were key descriptors that correlated with the dH_m and that the Gradient Boosting, Random Forest, and Extra Trees models achieved high predictive accuracy. The identification of fr_uncrch_alkane as an important descriptor highlighted the significance of alkane-like fragments in determining the stability of organic semiconductors. These findings had significant implications for the design and development of new organic semiconductor materials with improved stability and performance. The use of ML models to predict the dH_m of organic semiconductors accelerated the discovery of new materials and reduced the need for experimental synthesis and characterization. This approach was integrated with high-throughput computational screening and experimental validation to identify novel organic semiconductors with tailored properties. The development of more accurate and interpretable ML models provided deeper insights into the relationships between a molecular structure and material properties, enabling the design of organic semiconductors with improved efficiency, stability, and sustainability. Further studies also explored the application of ML models to predict other key properties of organic semiconductors, such as charge carrier mobility and optical absorption, to further accelerate the development of these materials for a wide range of applications, including solar cells, light-emitting diodes, and field-effect transistors.

Supplementary Materials: The following supporting information can be downloaded at https://www.mdpi.com/article/10.3390/coatings15070758/s1.

Author Contributions: S.N., S.H.S. and M.J.A. contributed equally to this work, sharing responsibility for conceptualization, formal analysis, methodology, software, validation, investigation, writing of the original draft, and review and editing of the manuscript. All authors have read and agreed to the published version of the manuscript.

Funding: This work was supported and funded by the Deanship of Scientific Research at Imam Mohammad Ibn Saud Islamic University (IMSIU) (grant number IMSIU-DDRSP2502).

Institutional Review Board Statement: Not applicable.

Informed Consent Statement: Not applicable.

Data Availability Statement: The calculated descriptors have been provided in the form of an Excel file, while their SMILES notation has been provided in the Supplementary File S1.

Conflicts of Interest: The authors declare no conflicts of interest.

References

1. Gopalakrishnan, V.; Balaji, D.; Dangate, M.S. Conjugated Polymer Photovoltaic Materials: Performance and Applications of Organic Semiconductors in Photovoltaics. *ECS J. Solid State Sci. Technol.* **2022**, *11*, 35001. [CrossRef]
2. Güleryüz, C.; Sumrra, S.H.; Hassan, A.U.; Mohyuddin, A.; Waheeb, A.S.; Awad, M.A.; Jalfan, A.R.; Noreen, S.; Kyhoiesh, H.A.K.; El Azab, I.H. A machine learning and DFT assisted analysis of benzodithiophene based organic dyes for possible photovoltaic applications. *J. Photochem. Photobiol. A Chem.* **2025**, *460*, 116157. [CrossRef]
3. Wu, T.; Tan, L.; Feng, Y.; Zheng, L.; Li, Y.; Sun, S.; Liu, S.; Cao, J.; Yu, Z. Toward Ultrathin: Advances in Solution-Processed Organic Semiconductor Transistors. *ACS Appl. Mater. Interfaces* **2024**, *16*, 61530–61550. [CrossRef]
4. Jia, Y.; Chen, G.; Zhao, L. Defect detection of photovoltaic modules based on improved VarifocalNet. *Sci. Rep.* **2024**, *14*, 15170. [CrossRef]
5. Dong, J.; Yan, C.; Chen, Y.; Zhou, W.; Peng, Y.; Zhang, Y.; Wang, L.-N.; Huang, Z.-H. Organic semiconductor nanostructures: Optoelectronic properties, modification strategies, and photocatalytic applications. *J. Mater. Sci. Technol.* **2022**, *113*, 175–198. [CrossRef]
6. Salzmann, I.; Heimel, G.; Oehzelt, M.; Winkler, S.; Koch, N. Molecular Electrical Doping of Organic Semiconductors: Fundamental Mechanisms and Emerging Dopant Design Rules. *Acc. Chem. Res.* **2016**, *49*, 370–378. [CrossRef] [PubMed]
7. Joly, D.; Delgado, J.L.; Atienza, C.; Martín, N. Light-Harvesting Materials for Organic Electronics. In *Photonics, Volume 2: Nanophotonic Structures and Materials*; Wiley: Hoboken, NJ, USA, 2015; pp. 311–341, ISBN 978-1-119-01401-0.
8. Weis, M. Organic Semiconducting Polymers in Photonic Devices: From Fundamental Properties to Emerging Applications. *Appl. Sci.* **2025**, *15*, 4028. [CrossRef]
9. Güleryüz, C.; Hassan, A.U.; Güleryüz, H.; Kyhoiesh, H.A.K.; Mahmoud, M.H.H. A machine learning assisted designing and chemical space generation of benzophenone based organic semiconductors with low lying LUMO energies. *Mater. Sci. Eng. B* **2025**, *317*, 118212. [CrossRef]
10. Kunkel, C.; Margraf, J.T.; Chen, K.; Oberhofer, H.; Reuter, K. Active discovery of organic semiconductors. *Nat. Commun.* **2021**, *12*, 2422. [CrossRef]
11. Wang, X.; Wang, W.; Yang, C.; Han, D.; Fan, H.; Zhang, J. Thermal transport in organic semiconductors. *J. Appl. Phys.* **2021**, *130*, 170902. [CrossRef]
12. Wang, X.; Peng, B.; Chan, P. Thermal Annealing Effect on the Thermal and Electrical Properties of Organic Semiconductor Thin Films. *MRS Adv.* **2016**, *1*, 1637–1643. [CrossRef]

13. Tong, W.; Li, H.; Liu, D.; Wu, Y.; Xu, M.; Wang, K. Study on the changes in the reverse recovery characteristics of high-power thyristor under 14.1 MeV fusion neutron irradiation. *Fusion Eng. Des.* **2025**, *211*, 114744. [CrossRef]
14. Miao, Z.; Gao, C.; Shen, M.; Wang, P.; Gao, H.; Wei, J.; Deng, J.; Liu, D.; Qin, Z.; Wang, P.; et al. Organic light-emitting transistors with high efficiency and narrow emission originating from intrinsic multiple-order microcavities. *Nat. Mater.* **2025**, *24*, 917–924. [CrossRef]
15. Güleryüz, C.; Sumrra, S.H.; Hassan, A.U.; Mohyuddin, A.; Elnaggar, A.Y.; Noreen, S. A machine learning analysis to predict the stability driven structural correlations of selenium-based compounds as surface enhanced materials. *Mater. Chem. Phys.* **2025**, *339*, 130786. [CrossRef]
16. Bronstein, H.; Nielsen, C.B.; Schroeder, B.C.; McCulloch, I. The role of chemical design in the performance of organic semiconductors. *Nat. Rev. Chem.* **2020**, *4*, 66–77. [CrossRef] [PubMed]
17. Wu, W.; Chen, Y.; Xie, B.; Wu, H.; Cheng, L.; Guo, Y.; Cai, C.; Chen, X. Microdynamic behaviors of Au/Ni-assisted chemical etching in fabricating silicon nanostructures. *App. Surf. Sci.* **2025**, *696*, 122915. [CrossRef]
18. Gao, S.; Wang, H.; Huang, H.; Dong, Z.; Kang, R. Predictive models for the surface roughness and subsurface damage depth of semiconductor materials in precision grinding. *IJEM* **2025**, *7*, 035103. [CrossRef]
19. Tian, X.; Xun, R.; Chang, T.; Yu, J. Distribution function of thermal ripples in h-BN, graphene and MoS_2. *Phys. Lett. A* **2025**, *550*, 130597. [CrossRef]
20. Wang, H.; Hou, Y.; He, Y.; Wen, C.; Giron-Palomares, B.; Duan, Y.; Gao, B.; Vavilov, V.P.; Wang, Y. A Physical-Constrained Decomposition Method of Infrared Thermography: Pseudo Restored Heat Flux Approach Based on Ensemble Bayesian Variance Tensor Fraction. *IEEE Trans. Ind. Inform.* **2023**, *20*, 3413–3424. Available online: https://ieeexplore.ieee.org/document/10242241 (accessed on 14 June 2025). [CrossRef]
21. Armeli, G.; Peters, J.-H.; Koop, T. Machine-Learning-Based Prediction of the Glass Transition Temperature of Organic Compounds Using Experimental Data. *ACS Omega* **2023**, *8*, 12298–12309. [CrossRef]
22. Qin, X.; Wang, Q.; Zhao, X.; Xia, S.; Wang, L.; Zhang, Y.; He, C.; Chen, D.; Jiang, B. PCS: Property-composition-structure chain in Mg-Nd alloys through integrating sigmoid fitting and conditional generative adversarial network modeling. *Scr. Mater.* **2025**, *265*, 116762. [CrossRef]
23. Mckinney, W. Pandas: A Foundational Python Library for Data Analysis and Statistics. *Python High Perform. Sci. Comput.* **2011**, *14*, 1–9.
24. Schäfer, C. Extensions for Scientists: NumPy, SciPy, Matplotlib, Pandas. In *Quickstart Python: An Introduction to Programming for STEM Students*; Schäfer, C., Ed.; Springer Fachmedien: Wiesbaden, Germany, 2021; pp. 45–53, ISBN 978-3-658-33552-6.
25. Scalfani, V.F.; Patel, V.D.; Fernandez, A.M. Visualizing chemical space networks with RDKit and NetworkX. *J. Cheminform.* **2022**, *14*, 87. [CrossRef] [PubMed]
26. Pedregosa, F.; Varoquaux, G.; Gramfort, A.; Michel, V.; Thirion, B.; Grisel, O.; Blondel, M.; Müller, A.; Nothman, J.; Louppe, G.; et al. Scikit-learn: Machine Learning in Python. *arXiv* **2018**, arXiv:1201.0490. [CrossRef]
27. Hasan, D.M.; Mallah, S.H.; Waheeb, A.S.; Güleryüz, C.; Hassan, A.U.; Kyhoiesh, H.A.K.; Elnaggar, A.Y.; Azab, I.H.E.; Mahmoud, M.H.H. Chemical modification-induced enhancements in quantum dot photovoltaics: A theoretical and molecular descriptive analysis. *Struct. Chem.* **2025**. [CrossRef]
28. Li, X.H.; Jalbout, A.F.; Solimannejad, M. Definition and application of a novel valence molecular connectivity index. *J. Mol. Struct. Theochem* **2003**, *663*, 81–85. [CrossRef]
29. Müller, M.; Hansen, A.; Grimme, S. An atom-in-molecule adaptive polarized valence single-ζ atomic orbital basis for electronic structure calculations. *J. Chem. Phys.* **2023**, *159*, 164108. [CrossRef]
30. Okoye, K.; Hosseini, S. Correlation Tests in R: Pearson Cor, Kendall's Tau, and Spearman's Rho. In *R Programming: Statistical Data Analysis in Research*; Okoye, K., Hosseini, S., Eds.; Springer Nature: Singapore, 2024; pp. 247–277, ISBN 978-981-97-3385-9.
31. Saarela, M.; Jauhiainen, S. Comparison of feature importance measures as explanations for classification models. *SN Appl. Sci.* **2021**, *3*, 272. Available online: https://link.springer.com/article/10.1007/s42452-021-04148-9 (accessed on 25 December 2024). [CrossRef]
32. Li, Q.; Yang, Y.; Wen, Y.; Tian, X.; Li, Y.; Xiang, W. A Fast Overcurrent Protection IC for SiC MOSFET Based on Current Detection. *IEEE Trans. Power Electron.* **2024**, *39*, 4986–4990. [CrossRef]
33. Hu, Q.-N.; Liang, Y.-Z.; Yin, H.; Peng, X.-L.; Fang, K.-T. Structural Interpretation of the Topological Index. 2. The Molecular Connectivity Index, the Kappa Index, and the Atom-type E-State Index. *J. Chem. Inf. Comput. Sci.* **2004**, *44*, 1193–1201. [CrossRef]
34. Natekin, A.; Knoll, A. Gradient boosting machines, a tutorial. *Front. Neurorobot.* **2013**, *7*, 21. [CrossRef] [PubMed]
35. Marques Ramos, A.P.; Prado Osco, L.; Elis Garcia Furuya, D.; Nunes Gonçalves, W.; Cordeiro Santana, D.; Pereira Ribeiro Teodoro, L.; Antonio da Silva Junior, C.; Fernando Capristo-Silva, G.; Li, J.; Henrique Rojo Baio, F.; et al. A random forest ranking approach to predict yield in maize with uav-based vegetation spectral indices. *Comput. Electron. Agric.* **2020**, *178*, 105791. [CrossRef]

36. Berrouachedi, A.; Jaziri, R.; Bernard, G. Deep Extremely Randomized Trees. In *Neural Information, Proceedings of the Neur Information Processing, Sydney, NSW, Australia, December 12–15, 2019*; Gedeon, T., Wong, K.W., Lee, M., Eds.; Springer Internationa Publishing: Cham, Switzerland, 2019; pp. 717–729.
37. Bentéjac, C.; Csörgő, A.; Martínez-Muñoz, G. A comparative analysis of gradient boosting algorithms. *Artif. Intell. Rev.* **202** *54*, 1937–1967. [CrossRef]
38. Fox, E.W.; Hill, R.A.; Leibowitz, S.G.; Olsen, A.R.; Thornbrugh, D.J.; Weber, M.H. Assessing the accuracy and stability of variabl selection methods for random forest modeling in ecology. *Environ. Monit. Assess.* **2017**, *189*, 316. [CrossRef]
39. Mallah, S.H.; Güleryüz, C.; Sumrra, S.H.; Hassan, A.U.; Güleryüz, H.; Mohyuddin, A.; Kyhoiesh, H.A.K.; Noreen, S.; Elnagga A.Y. Benzothiophene semiconductor polymer design by machine learning with low exciton binding energy: A vast chemica space generation for new structures. *Mater. Sci. Semicond. Process.* **2025**, *190*, 109331. [CrossRef]
40. Liebscher, E. Estimating the Density of the Residuals in Autoregressive Models. *Stat. Inference Stoch. Process.* **1999**, *2*, 105–11 [CrossRef]
41. Zhou, Y.; Fan, S.; Zhu, Z.; Su, S.; Hou, D.; Zhang, H.; Cao, Y. Enabling High-Sensitivity Calorimetric Flow Sensor Using Vanadiun Dioxide Phase-Change Material with Predictable Hysteretic Behavior. *IEEE Trans. Electron. Devices* **2025**, *72*, 1360–1367. [CrossRef
42. Zhang, X.; Liu, C.-A. Model averaging prediction by *K*-fold cross-validation. *J. Econom.* **2023**, *235*, 280–301. [CrossRef]
43. Young, S.R.; Rose, D.C.; Karnowski, T.P.; Lim, S.-H.; Patton, R.M. Optimizing deep learning hyper-parameters through a evolutionary algorithm. In Proceedings of the Workshop on Machine Learning in High-Performance Computing Environment Austin, TX, USA, 15 November 2015; Association for Computing Machinery: New York, NY, USA, 2015; pp. 1–5.
44. Cess, C.G.; Haghverdi, L. Compound-SNE: Comparative alignment of t-SNEs for multiple single-cell omics data visualizatior *Bioinformatics* **2024**, *40*, btae471. [CrossRef]

coatings

Article

Influence of Ce^{3+} Doping on Photoluminescence Properties and Stability of Cs_4SnBr_6 Zero-Dimensional Perovskite

Xinye Lu, Haixia Wu, Jisheng Xu, Jianni Chen, Yaqian Huang, Hongliang Li *, Jie Song and Rui Huang *

School of Materials Science and Engineering, Hanshan Normal University, Chaozhou 521041, China; xinyelu2023@hotmail.com (X.L.); 20220063@hstc.edu.cn (H.W.); jishengxu2023@hotmail.com (J.X.); chenjianni0130@hotmail.com (J.C.); yaqianhuang1208@hotmail.com (Y.H.); songjie@hstc.edu.cn (J.S.)
* Correspondence: lhl4@hstc.edu.cn (H.L.); rhuang@hstc.edu.cn (R.H.)

Abstract: Zero-dimensional tin-based halide perovskites have garnered considerable interest owing to their remarkable optical properties, including broad-band emission, high photoluminescence (PL) efficiency, and low self-absorption. Nevertheless, enhancing the PL efficiency and stability of these materials remains a pressing challenge. In this study, the enhancement of PL and stability in Cs_4SnBr_6 zero-dimensional perovskite was investigated through Ce^{3+} doping. Our experimental results demonstrate that the incorporation of Ce^{3+} can significantly boost the light emission intensity from self-trapped excitons (STEs) in Cs_4SnBr_6, achieving over a 150% increase compared to the undoped sample, with a PL quantum yield of approximately 64.7%. Moreover, the thermal stability of the corresponding doped sample is markedly enhanced. Through comprehensive analyses, including X-ray diffraction, energy-dispersive spectroscopy, time-resolved PL, and temperature-dependent PL measurements, we elucidate that the enhanced light emission is attributed to the distortion of the $[SnBr_6]^{4-}$ octahedral structure induced by Ce^{3+} doping, which strengthens electron–phonon coupling and elevates the binding energy of STEs.

Keywords: photoluminescence; self-trapped exciton; Cs_4SnBr_6; Ce^{3+} doping

1. Introduction

In recent years, metal halide perovskites have attracted considerable interest owing to their unique optoelectronic properties, such as high photoluminescence quantum yields, high carrier mobility, and substantial absorption coefficients. These semiconductors have demonstrated remarkable performance in photovoltaics, light-emitting devices, photodetectors, and beyond [1–4]. One of the most successful applications of three-dimensional halide perovskites is as the light-emitting layer in light-emitting devices [5,6]. The efficiency of these devices has dramatically increased from 8.5% in 2015 to over 30% in recent years [5,6]. However, the high toxicity of lead poses a significant threat to human health and the environment, limiting its commercial applications. Consequently, researchers have explored non-toxic or low-toxic alternatives, such as Sn^{2+} and Ge^{2+} equivalent substitutes and Bi^{3+} and Ag^{+} alternative substitutes, to develop lead-free perovskite luminescent materials [7–10]. Among these alternative elements, Sn^{2+} is considered the most promising substitute due to its similar coordination number and charge balance characteristics compared with Pb^{2+}. Furthermore, the ionic radius of Sn closely matches that of Sn, which helps avoid significant lattice distortion caused by substitution. In addition, compared to lead-based perovskites, tin-based perovskites exhibit narrower band gaps and higher charge carrier mobility. Most importantly, the degradation product SnO_2 formed by tin-based perovskites upon exposure to environmental air is an environmentally friendly material that does not pollute the environment [11]. Therefore, the use of low-toxicity tin to replace lead in halide perovskites ($CsPbX_3$) is of great interest. To date, considerable efforts have been made to explore efficient and stable $CsSnX_3$ [12–14]. For instance, Liu et al. proposed a colloid synthesis strategy that successfully yielded narrow band-emitting

$CsSnI_3$ nanocrystals with an emission efficiency of up to 18.4% by regulating the reactant ratios [12]. Similarly, Kang et al. developed a method for synthesizing $CsSnX_3$ nanocrystals using Cs_2CO_3, SnC_2O_4, and NH_4X as raw materials, achieving the stability of $CsSnX_3$ nanocrystals by employing the antioxidant SnC_2O_4 as the reaction precursor [13]. Although $CsSnX_3$ perovskites have made significant progress, their luminescent efficiency and stability still fall short of meeting the requirements for commercial applications.

Compared to three-dimensional cesium lead halide perovskites ($CsSnX_3$), zero-dimensional cesium tin halide perovskites (Cs_4SnBr_6) have received widespread interest owing to their higher stability and PL quantum yield [15–17]. In zero-dimensional cesium tin halide perovskites, the strong electron–phonon interaction arising from their soft lattice characteristics facilitates the formation of states with self-trapped excitons. This results in efficient emission across the visible spectrum, making them ideal candidates for optoelectronic devices such as white LEDs [15,16]. For example, Kovalenko et al. demonstrated strong light emission from self-trapped excitons in Cs_4SnBr_6, with a room-temperature PL quantum yield reaching $15 \pm 5\%$ [15]. Up to now, significant efforts have been made to enhance the emission efficiency and stability of zero-dimensional halide perovskites [17–20]. For example, to address the air instability of Cs_4SnBr_6, Zhang and his colleagues substituted SnF_2 for the easily oxidized $SnBr_2$ as the tin source, effectively enhancing the structural stability of Cs_4SnBr_6 by utilizing F to suppress Sn^{2+} oxidation [17]. On the other hand, Ma and his team demonstrated that high pressure can robustly boost the PL efficiency of Cs_4SnBr_6, which is attributed to the amplified optical activity and increased binding energy of self-trapped excitons under compression [18]. In previous work, we extended the emission spectrum of Cs_4SnBr_6 through an Mn^{2+} doping strategy, enhancing the electron–phonon coupling effect as well as the binding energy of STEs, thereby significantly increasing the emission efficiency of Cs_4SnBr_6. The PL quantum yield reached approximately 75%, and the thermal stability was also improved [20]. In addition, we proposed a new strategy using rapid thermal treatment (RTT) to improve the emission of STEs in Cs_4SnBr_6. The improved emission is ascribed to the suitable electron–phonon coupling as well as the augmented binding energies of STEs brought about by RTT [21]. In zero-dimensional perovskites, the introduction of ions with different sizes at the B site can lead to deviations in the ionic radius, causing distortion of the $[BX_6]^{4-}$ octahedral structure. This subtle structural change may open new avenues for controlling STE emissions. It is well known that the ionic radius of Ce^{3+} (1.01 Å) is slightly smaller than that of Sn^{2+} (1.18 Å) in a similar coordination environment. This indicates that substituting Ce^{3+} for Sn^{2+} in a crystal lattice can induce distortion in the $[BX_6]^{4-}$ octahedral structure. However, the effects of Ce^{3+} doping on the PL characteristics and stability of Cs_4SnBr_6 remain unexplored to date.

In this paper, the synthesis of Ce^{3+}-doped Cs_4SnBr_6 zero-dimensional perovskites was achieved via a water-assisted ball milling method, and the impact of Ce^{3+} doping on the luminescence characteristics and stability of Cs_4SnBr_6 were investigated. We demonstrated that the incorporation of Ce^{3+} significantly could boost the light emission intensity from STEs in Cs_4SnBr_6, achieving over a 150% increase compared to the undoped sample, with a PL quantum yield (QY) of approximately 64.74%. Moreover, the thermal stability of the corresponding doped sample is markedly enhanced. Through comprehensive analyses, including X-ray diffraction, energy-dispersive spectroscopy, time-resolved PL, and temperature-dependent PL measurements, the origin of the improved light emission was discussed.

2. Materials and Methods

2.1. Materials

Cesium bromide (CsBr, 99.9%), ammonium bromide (NH_4Br, 99.99%), and cerium(III) bromide ($CeBr_3$, 99.9%) were procured from Aladdin (Meridian Parkway, Riverside, CA, USA). Stannous fluoride (SnF_2, 99.9%) was obtained from Macklin (Shanghai Macklin Biochemical Co., Ltd., Shanghai, China).

2.2. Synthesis of Cs_4SnBr_6 and Ce^{3+}-Doped Cs_4SnBr_6

Cs_4SnBr_6 and Ce^{3+}-doped Cs_4SnBr_6 powders were synthesized via a water-assisted wet ball-milling procedure, respectively. The reactant precursors employed in this study were CsBr, SnF_2, NH_4Br, and $CeBr_3$. To synthesize Ce^{3+}-doped Cs_4SnBr_6 with different levels, the molar quantities of CsBr, SnF_2, and NH_4Br were kept at 4 mmol, 1 mmol, and 2 mmol, respectively, and the molar quantities of $CeBr_3$ were kept at 0 mmol, 0.1 mmol, 0.3 mmol, 0.5 mmol, and 0.8 mmol, corresponding to the samples named S-x (x = 0, 0.1, 0.3, 0.5, 0.8). The experimental procedure is as follows: the required precursors were thoroughly mixed and placed into a ball milling jar, to which 60 μL of deionized water was added. The mixture was then subjected to ball milling at 600 rpm for 30 min. After milling, the resulting product was subjected to vacuum drying by placing it in an oven set at 60 °C for a duration of 4 h. The sample was subsequently cooled to room temperature and subjected to additional dry ball milling at the same speed for 30 min. This process yielded the desired Ce^{3+}-doped Cs_4SnBr_6 powder sample.

2.3. Characterizations

The PL properties were meticulously assessed by an Edinburgh Instrument FLS1000 PL spectrometer (Livingstone, UK). To examine the temperature-dependent behavior, PL spectra were recorded across a range of temperatures. Additionally, PL excitation spectra and PL decay were captured by utilizing the same instrument. The crystallographic structures of samples were examined through X-ray diffraction (XRD) with an X'pertPro Panlytical diffractometer (PANalytical, Almelo, The Netherlands), operating at 35 kV and 40 mA with Cu Kα (0.15418 nm) filtered radiation. The surface morphology and compositional analysis of samples were investigated via scanning electron microscopy (SEM) using a Hitachi SU5000 SEM instrument (Tokyo, Japan) and a Bruker EDS QUANTAX system (Karlsruhe, Germany), respectively.

3. Results and Discussion

Figure 1a presents the SEM surface morphology of the S-0.3 sample. Figure 1b depicts the qualitative analysis spectrum acquired through EDS spectra for Br, Sn, Cs, F, and Ce elements in the S-0.3 sample. The distribution maps for Br, Sn, Cs, F, and Ce elements, respectively, are also shown in Figure 1c–g, which clearly indicate a relatively uniform distribution of these elements. Figure 2 illustrates the variation in the content of Ce, Sn, and Br elements across samples with various Ce^{3+} doping concentrations. The data presented in Figure 2 demonstrate a negative correlation between the Ce^{3+} doping concentration and the Sn content, indicating that an increase in Ce^{3+} doping leads to a corresponding decrease in Sn content. This suggests that Ce^{3+} ions are incorporated into the Cs_4SnBr_6 lattice, substituting for some of the Sn^{2+} sites.

Figure 3 presents the XRD patterns of samples with varying Ce^{3+} doping concentrations. The diffraction peaks correspond to the (110), (131), (223), (300), and (330) crystal planes of the Cs_4SnBr_6 phase, in agreement with previously reported patterns for SnF_2-derived Cs_4SnBr_6 [17,22]. The high crystallinity of the samples is evidenced by these discernible diffraction peaks, suggesting that the primary crystal structure of Cs_4SnBr_6 remains undisturbed even with Ce^{3+} doping. Moreover, increasing the Ce^{3+} doping concentration results in the gradual deviation of the diffraction peaks from the standard positions, shifting towards higher 2θ values. This shift is attributed to lattice shrinkage caused by the substitution of Sn^{2+} (r = 1.18 Å, coordination number = 6) with the smaller Ce^{3+} ions (r = 1.01 Å, coordination number = 6) in the octahedral sites. Additionally, a strong diffraction peak at 29.7° was observed, originating from the CsBr phase. This indicates an incomplete chemical reaction involving the CsBr powder precursors during the solid-state synthesis process.

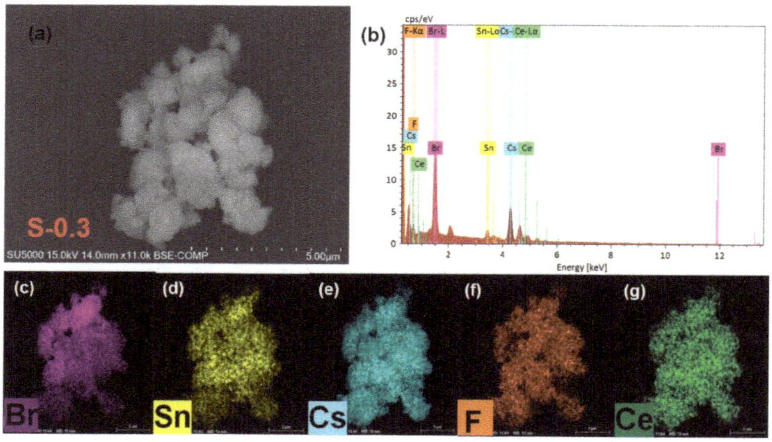

Figure 1. (**a**) The surface morphology image of the S-0.3 sample. (**b**) The signal intensity relationships of different elements. (**c**–**g**) The elemental distribution maps of Br, Sn, Cs, F, and Ce in the S-0.3 sample.

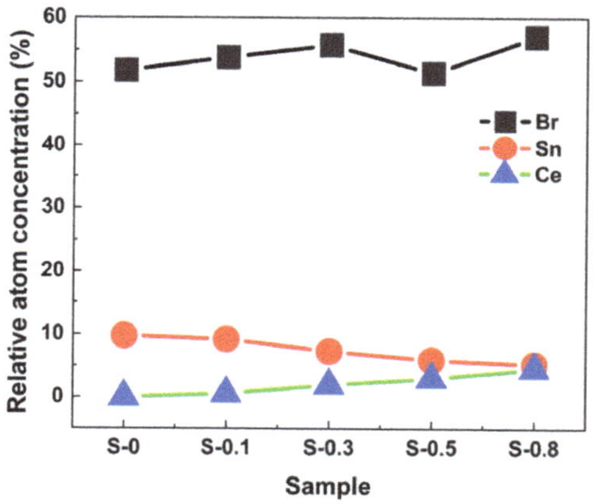

Figure 2. The Br, Sn, and Ce elemental content in Cs_4SnBr_6 samples with different Ce^{3+} doping levels.

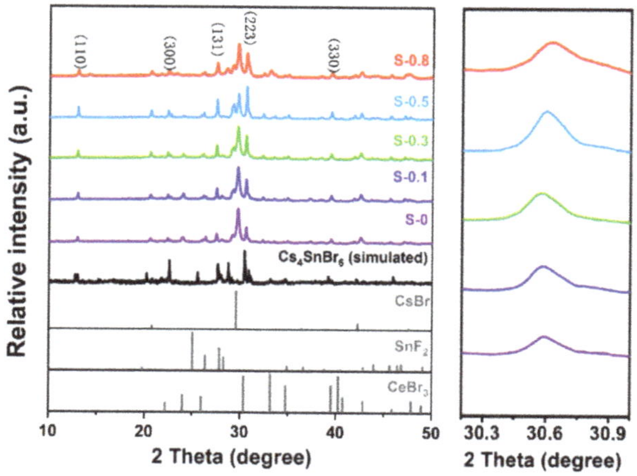

Figure 3. XRD patterns of Cs_4SnBr_6 powders with various Ce^{3+} doping levels, respectively.

Figure 4 illustrates the PL characteristics of samples with varying Ce^{3+} doping concentrations. As shown in Figure 4a, the PL intensity increases progressively with higher Ce^{3+} doping concentrations, peaking at 525 nm with the maximum green emission when the Ce^{3+} doping concentration reaches 0.3 mmol. At this concentration, the PL efficiency attains approximately 64.74%, as depicted in Figure 4c. Beyond this point, further increases in Ce^{3+} doping concentration result in a decrease in PL intensity, likely due to the formation of non-radiative recombination centers caused by excessive Ce^{3+} doping, which in turn diminishes the PL intensity. Moreover, Figure 4b indicates that introducing a small amount of Ce^{3+} significantly enhances the PLE intensity from Cs_4SnBr_6 samples. This enhancement is beneficial for improving light absorption and conversion, thereby boosting light emission. A comparison of Figure 4a,b reveals that all samples exhibit a substantial Stokes shift of approximately 1.3 eV.

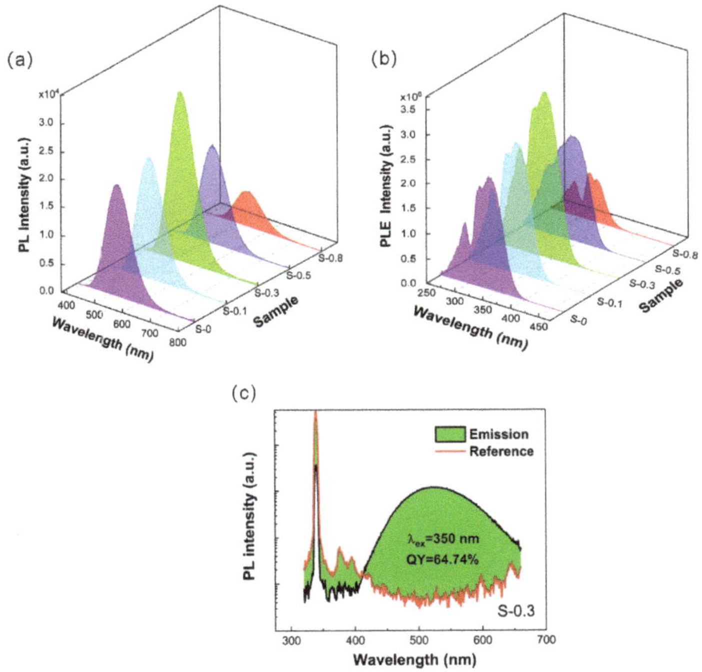

Figure 4. (**a**) PL intensity of Cs_4SnBr_6 samples with different Ce^{3+} doping levels. The PL spectra are excited by the 340 nm line from the Xe lamp. (**b**) PLE intensity of Cs_4SnBr_6 samples with different Ce^{3+} doping levels. The PLE spectra are monitored at 525 nm. (**c**) The PL QY was acquired from the PLE and PL spectra for the Cs_4SnBr_6 sample 0.3.

To better understand the PL properties, we measured the PL decay curves using an excitation wavelength of 375 nm, facilitated by 70 ps laser pulses. As depicted in Figure 5, the green emission from all samples exhibits a slow decay characteristic. We observed that all PL decay curves can be well-fitted by a biexponential function, described by the equation [23]:

$$I(t) = A_1 e^{-t/\tau_1} + A_2 e^{-t/\tau_2} \quad (1)$$

where τ_1 and τ_2 are the time constants, and A_1 and A_2 are the normalized amplitudes of the lifetime components. The intensity-weighted average PL lifetime was determined by $(A_1 * \tau_1^2 + A_2 * \tau_2^2)/(A_1 * \tau_1 + A_2 * \tau_2)$ accordingly [23]. Notably, the PL lifetimes of all samples are on the order of microseconds, indicative of self-trapped exciton emission characteristics [20,22]. However, we observed a significant decrease in PL lifetime as the

$CeBr_3$ molar ratio increased to 0.8 mmol, suggesting that excessive Ce^{3+} doping increases the probability of non-radiative recombination. Combined with the XRD analysis in Figure 3, we inferred that high Ce^{3+} doping levels cause severe distortion of the $[SnBr_6]^{4-}$ octahedra in Cs_4SnBr_6, resulting in the formation of more defect states. This finding aligns with the observed decrease in emission intensity at higher Ce^{3+} doping levels, as shown in Figure 4a.

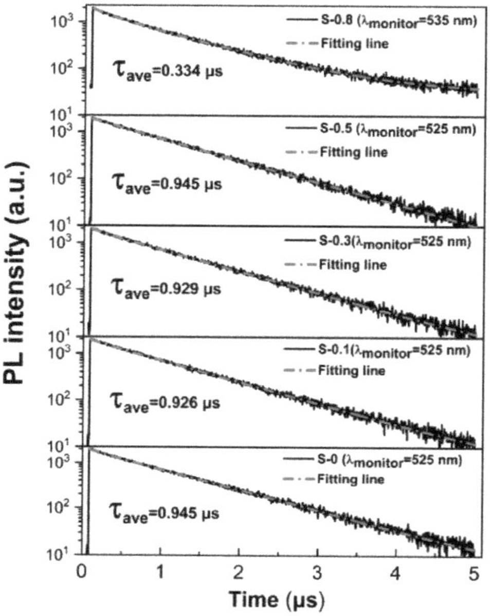

Figure 5. Time-resolved PL decay traces were recorded at 525 nm in Ce^{3+}-doped Cs_4SnBr_6 perovskites with different doping levels. Each measurement was performed using an excitation wavelength of 375 nm with 70 ps laser pulses.

Figure 6 illustrates the relationship between PL intensity and pumping power for the S-0.3 sample. As depicted in the inset, the PL intensity increases consistently as the pumped power rises from 160.3 nW to 2817 nW. Notably, the PL peak position remains constant, unaffected by changes in pumped power. It is well known that the dependence of PL intensity on excitation power can be used to infer the underlying mechanisms of light emission in semiconductors. The PL intensity I can be expressed by the equation [24]:

$$I = \eta I_0^k \tag{2}$$

where I_0 represents the pumped power, η is the emission efficiency, and k is an exponent related to the radiative recombination process. By performing a linear fit of $\ln(I/\eta)$ versus $\ln(I_0)$, we can determine the value of k. As shown in Figure 6, there is a clear linear relationship between the integrated PL intensity and the pumped power in the range of 160.3 to 2817 nW, yielding a k value of 1.11. This suggests that the green emission from the S-0.3 sample can be attributed to self-trapped exciton states [20,24].

Figure 7a shows the PL intensity variation for the S-0.3 sample across a temperature range of 100–300 K. As observed, the PL intensity decreases with the rise in temperature, primarily due to thermal activation causing the detrapping of STEs, leading to a reduction in the radiative recombination rate. The light emission efficiency of STEs is greatly influenced by the exciton-binding energy. According to the Arrhenius equation, the relationship between the integrated PL intensity I_{PL} (T) at different temperatures and the exciton binding energy of STEs can be expressed as [20]:

$$I_{PL}(T) = \frac{I_{PL}(T_0)}{1 + \beta \exp(-E_b/k_B T)} \quad (3)$$

where $I_{PL}(T_0)$ represents the integrated PL intensity at 100 K, β is a constant associated with radiative recombination centers' density, K_B denotes the Boltzmann constant, and E_b signifies the exciton binding energy. By fitting the data to Equation (3) of the Arrhenius equation, the exciton binding energy for the S-0.3 sample was determined to be 494 meV, significantly higher than that of the S-0 sample (265 meV) [20]. This indicates that Ce^{3+} doping can effectively suppress the thermal activation-induced detrapping of STEs in Cs_4SnBr_6, thereby enhancing efficient emission from STEs.

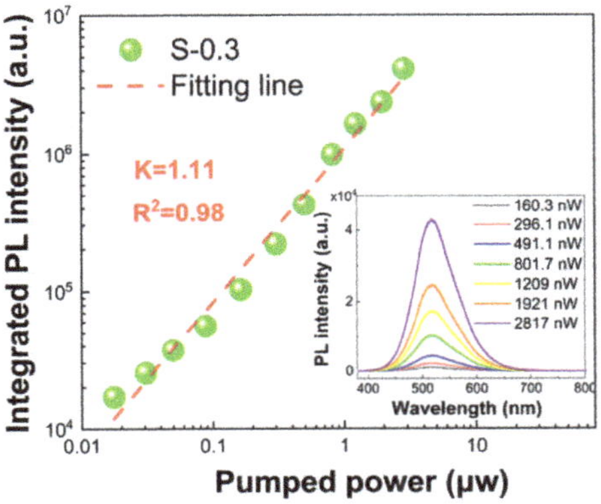

Figure 6. Integrated PL intensity as a function of pumped power for the S-0.3 sample. The red dashed line represents the fitting curve, yielding a power–law exponent (K) of 1.11 with a coefficient of determination (R^2) of 0.98. The inset shows the PL spectra of the S-0.3 sample under various pumped powers.

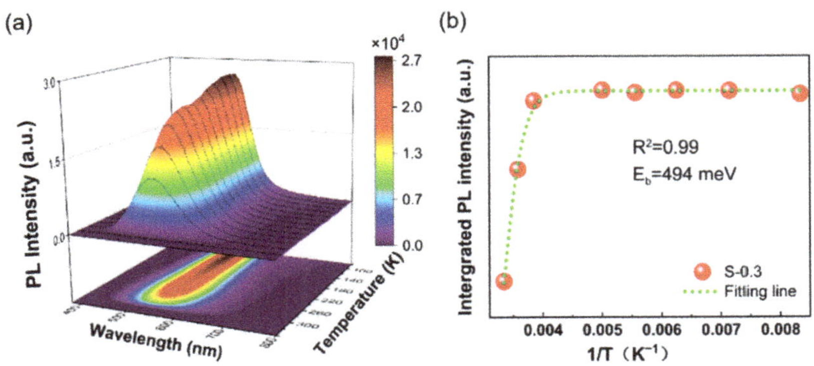

Figure 7. (**a**) Temperature-dependent PL spectra of the S-0.3 sample. The 3D plot displays PL intensity as a function of wavelength and temperature. The color scale indicates the PL intensity in arbitrary units (a.u.), showing a peak in the lower temperature range. (**b**) Arrhenius plot of the integrated PL intensity versus inverse temperature (1/T) for the S-0.3 sample. The green dotted line represents the fitting curve, yielding an exciton binding energy E_b of 494 meV with a coefficient of determination R^2 of 0.99.

To further understand the self-trapped exciton states, the impact of electron–phonon coupling on the PL characteristics was investigated. According to Stadler's theory [25], the full width at half maximum (FWHM) of PL spectra is closely associated with electron–phonon coupling and can be mathematically described as follows:

$$\mathrm{FWHM(T)} = 2.36\sqrt{S}\mathrm{h}\omega\sqrt{\coth\frac{\mathrm{h}\omega}{2k_B T}} \tag{4}$$

where S represents the Huang–Rhys factor, hω denotes the optical phonon energy, and K_B signifies the Boltzmann constant. The Huang–Rhys factor S is widely employed to characterize the exciton-phonon coupling effect. As shown in Figure 8, the S value for the S-0.3 sample is 46.5, higher than that (41.2) of the undoped sample [20], indicating enhanced electron–phonon coupling in the Ce^{3+}-doped sample. This enhancement facilitates the formation of STEs. From the fit using Equation (4), the optical phonon energy E_{LO} is found to be 17.6 meV (141 cm^{-1}), which closely matches the Sn–Br stretching vibration mode observed around 128 cm^{-1} in Cs_4SnBr_6 (as shown in the inset of Figure 8) [26]. This indicates that the phonon modes involved in the electron–phonon coupling correspond to the Sn–Br stretching vibrations. Based on the above analysis and combined with the XRD results, we infer that Ce^{3+} doping in Cs_4SnBr_6 zero-dimensional perovskite induces structural distortions in the $[SnBr_6]^{4-}$ octahedra. These subtle structural changes enhance the electron–phonon coupling effect and increase the exciton binding energy, which are key factors for the observed enhancement of STE emission in the S-0.3 sample.

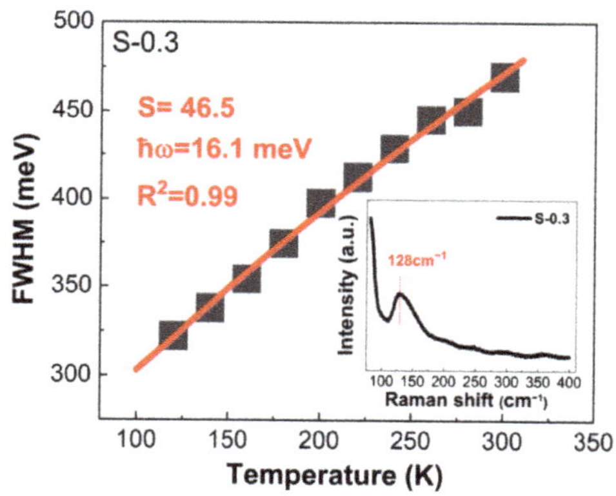

Figure 8. The FWHM of the PL spectra as a function of temperature for the S-0.3 sample. The red line represents the linear fit of the data, yielding a Huang–Rhys factor (S) of 46.5 and an optical phonon energy (hω) of 16.1 meV, with a coefficient of determination (R^2) of 0.99. The inset displays the Raman spectrum of the S-0.3 sample.

To evaluate the thermal stability of the S-0.3 sample, we monitored the changes in the integrated PL intensity at various temperatures through heating and cooling cycles. As illustrated in Figure 9, the S-0.3 sample exhibits PL thermal quenching as the temperature increases from 298 K to 398 K. Remarkably, upon cooling, the PL intensity nearly recovers to its original state, showing minimal reduction. In stark contrast, the S-0 sample demonstrates a significant decline in PL intensity, exceeding 90%, after undergoing the same thermal cycling [20]. These results clearly indicate that the S-0.3 sample has superior thermal stability compared to the S-0 sample.

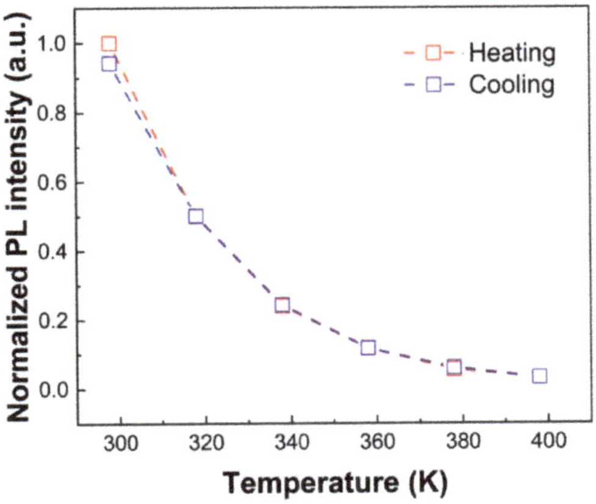

Figure 9. Heating and cooling cycle measurements of the S-0.3 sample across various temperatures.

4. Conclusions

In summary, Ce^{3+}-doped Cs_4SnBr_6 zero-dimensional perovskite powders were successfully prepared by utilizing a water-assisted wet ball-milling method. Our study demonstrates that Ce^{3+} doping significantly enhances the PL efficiency and thermal stability of Cs_4SnBr_6. The Ce^{3+}-doped samples exhibited a notable increase in light emission intensity from STEs, with an over 150% improvement compared to the undoped sample, achieving a PL QY of approximately 64.7%. Comprehensive analyses, including time-resolved PL, XRD, EDS, and temperature-dependent PL measurements, reveal that the enhanced PL results from the structural distortion of the $[SnBr_6]^{4-}$ octahedra induced by Ce^{3+} doping. This structural alteration strengthens electron–phonon coupling and increases the binding energy of STEs, thereby contributing to improved luminescence and thermal stability. These findings underscore the potential of Ce^{3+}-doped Cs_4SnBr_6 zero-dimensional tin-based perovskites for applications in optoelectronic devices, paving the way for future research to further optimize these materials for practical applications.

Author Contributions: X.L.: investigation, formal analysis, writing—original draft. H.W.: investigation, formal analysis. J.X.: investigation, formal analysis. Y.H.: investigation J.C.: investigation. J.S.: investigation. H.L.: formal analysis, writing—review and editing. R.H.: writing—review and editing, formal analysis, funding acquisition. All authors have read and agreed to the published version of the manuscript.

Funding: This research was funded by the Project of Guangdong Province Key Discipline Scientific Research Level Improvement (2021ZDJS039) and the Special Fund for Science and Technology Innovation Strategy of Guangdong Province (pdjh2023a0336).

Institutional Review Board Statement: Not applicable.

Informed Consent Statement: Not applicable.

Data Availability Statement: Data underlying the results presented in this paper are not publicly available at this time but may be obtained from the authors upon reasonable request.

Conflicts of Interest: The authors declare no conflicts of interest.

References

1. Liu, X.K.; Xu, W.; Bai, S.; Jin, Y.; Wang, J.; Friend, R.H.; Gao, F. Metal halide perovskites for light-emitting diodes. *Nat. Mater.* **2021**, *20*, 10–21. [CrossRef] [PubMed]
2. Hsu, W.-J.; Pettit, E.C.; Swartwout, R.; Zhitomirsky Kadosh, T.; Srinivasan, S.; Wassweiler, E.L.; Haugstad, G.; Bulović, V.; Holmes, R.J. Efficient Metal-Halide Perovskite Photovoltaic Cells Deposited via Vapor Transport Deposition. *Sol. RRL* **2024**, *8*, 2300758. [CrossRef]

3. Wu, S.; Yuan, L.; Chen, G.; Peng, C.; Jiang, Y. All-inorganic Mn^{2+}-doped metal halide perovskite crystals for the late-time detection of X-ray afterglow imaging. *Nanoscale* **2023**, *15*, 13628–13634. [CrossRef]
4. Girolami, M.; Matteocci, F.; Pettinato, S.; Serpente, V.; Bolli, E.; Paci, B.; Generosi, A.; Salvatori, S.; Di Carlo, A.; Trucchi, D.M. Metal-halide perovskite submicrometer-thick films for ultra-stable self-powered direct X-ray detectors. *Nano-Micro Lett.* **2024**, *16*, 182. [CrossRef] [PubMed]
5. Cho, H.; Jeong, S.H.; Park, M.H.; Kim, Y.H.; Wolf, C.; Lee, C.L.; Heo, J.H.; Sadhanala, A.; Myoung, N.; Yoo, S.; et al. Overcoming the electroluminescence efficiency limitations of perovskite light-emitting diodes. *Science* **2015**, *350*, 1222–1225. [CrossRef]
6. Bai, W.; Xuan, T.; Zhao, H.; Dong, H.; Cheng, X.; Wang, L.; He, R.J. Perovskite light-emitting diodes with an external quantum efficiency exceeding 30%. *Adv. Mater.* **2023**, *35*, 2302283. [CrossRef] [PubMed]
7. Fan, Q.; Biesold-McGee, G.V.; Ma, J.; Xu, Q.; Pan, S.; Peng, J.; Lin, Z. Lead-free halide perovskite nanocrystals: Crystal structures, synthesis, stabilities, and optical properties. *Angew. Chem. Int. Ed.* **2020**, *59*, 1030–1046. [CrossRef] [PubMed]
8. Arfin, H.; Nag, A. Origin of luminescence in Sb^{3+}- and Bi^{3+}-doped Cs_2SnCl_6 perovskites: Excited state relaxation and spin–orbit coupling. *J. Phys. Chem. Lett.* **2021**, *12*, 10002–10008. [CrossRef]
9. Infante, I.; Manna, L. Are there good alternatives to lead halide perovskite nanocrystals? *Nano Lett.* **2020**, *21*, 6–9. [CrossRef]
10. Xu, K.; Wei, Q.; Wang, H.; Yao, B.; Zhou, W.; Gao, R.; Chen, H.; Li, H.; Wang, T.; Ning, Z. The 3D-structure-mediated growth of zero-dimensional Cs_4SnX_6 nanocrystals. *Nanoscale* **2022**, *14*, 2248–2255. [CrossRef]
11. Ke, W.; Kanatzidis, M.G. Prospects for low-toxicity lead-free perovskite solar cells. *Nat. Commun.* **2019**, *10*, 965. [CrossRef] [PubMed]
12. Liu, Q.; Yin, J.; Zhang, B.B.; Chen, J.K.; Zhou, Y.; Zhang, L.M.; Wang, L.M.; Zhao, Q.; Hou, J.; Shu, J.; et al. Theory-guided synthesis of highly luminescent colloidal cesium tin halide perovskite nanocrystals. *J. Am. Chem. Soc.* **2021**, *143*, 5470–5480. [CrossRef]
13. Kang, C.; Rao, H.; Fang, Y.; Zeng, J.; Pan, Z.; Zhong, H. Antioxidative stannous oxalate derived lead-free stable $CsSnX_3$ (X = Cl, Br, and I) perovskite nanocrystals. *Angew. Chem. Int. Ed.* **2021**, *60*, 660–665. [CrossRef]
14. Zhang, B.B.; Chen, J.K.; Zhang, C.; Shirahata, N.; Sun, H.T. Mechanistic Insight into the Precursor Chemistry of Cesium Tin Iodide Perovskite Nanocrystals. *ACS Mater. Lett.* **2023**, *5*, 1954–1961. [CrossRef]
15. Benin, B.M.; Dirin, D.N.; Morad, V.; Wörle, M.; Yakunin, S.; Rainò, G.; Nazarenko, O.; Fischer, M.; Infante, I.; Kovalenko, M.V. Highly emissive self-trapped excitons in fully inorganic zero-dimensional tin halides. *Angew. Chem. Int. Ed.* **2018**, *57*, 11329–11333. [CrossRef] [PubMed]
16. Tan, L.; Wang, W.; Li, Q.; Luo, Z.; Zou, C.; Tang, M.; Zhang, L.; He, J.; Quan, Z. Colloidal syntheses of zero-dimensional Cs_4SnX_6 (X = Br, I) nanocrystals with high emission efficiencies. *Chem. Commun.* **2020**, *56*, 387–390. [CrossRef] [PubMed]
17. Zhang, Q.; Liu, S.; He, M.; Zheng, W.; Wan, Q.; Liu, M.; Liao, X.; Zhan, W.; Yuan, C.; Liu, J.; et al. Stable Lead-Free Tin Halide Perovskite with Operational Stability > 1200 h by Suppressing Tin (II) Oxidation. *Angew. Chem.* **2022**, *134*, e202205463. [CrossRef]
18. Ma, Z.; Liu, Z.; Lu, S.; Wang, L.; Feng, X.; Yang, D.; Wang, K.; Xiao, G.; Zhang, L.; Redfern, S.A.; et al. Pressure-induced emission of cesium lead halide perovskite nanocrystals. *Nat. Commun.* **2018**, *9*, 4506. [CrossRef]
19. Li, K.; Ye, Y.; Zhang, W.; Zhou, Y.; Zhang, Y.; Lin, S.; Lin, H.; Ruan, J.; Liu, C. Ultra-stable and color-tunable manganese ions doped lead-free cesium zinc halides nanocrystals in glasses for light-emitting applications. *Nano Res.* **2022**, *15*, 9368–9376. [CrossRef]
20. Lin, Z.; Wang, A.; Huang, R.; Wu, H.; Song, J.; Lin, Z.; Hou, D.; Zhang, Z.; Guo, Y.; Lan, S. Manipulating the sublattice distortion induced by Mn^{2+} doping for boosting the emission characteristics of self-trapped excitons in Cs_4SnBr_6. *J. Mater. Chem. C* **2023**, *11*, 5680–5687. [CrossRef]
21. Wu, H.; Lin, Z.; Song, J.; Zhang, Y.; Guo, Y.; Zhang, W.; Huang, R. Boosting the Self-Trapped Exciton Emission in Cs4SnBr6 Zero-Dimensional Perovskite via Rapid Heat Treatment. *Nanomaterials* **2023**, *13*, 2259. [CrossRef] [PubMed]
22. Chiara, R.; Ciftci, Y.O.; Queloz, V.I.; Nazeeruddin, M.K.; Grancini, G.; Malavasi, L. Green-emitting lead-free Cs_4SnBr_6 zero-dimensional perovskite nanocrystals with improved air stability. *J. Phys. Chem. Lett.* **2020**, *11*, 618–623. [CrossRef] [PubMed]
23. Lin, K.H.; Liou, S.C.; Chen, W.L.; Wu, C.L.; Lin, G.R.; Chang, Y.M. Tunable and stable UV-NIR photoluminescence from annealed SiO_x with Si nanoparticles. *Opt. Express* **2013**, *21*, 23416–23424. [CrossRef] [PubMed]
24. Bergman, L.; Chen, X.B.; Morrison, J.L.; Huso, J.; Purdy, A.P. Photoluminescence dynamics in ensembles of wide-band-gap nanocrystallites and powders. *J. Appl. Phys.* **2004**, *96*, 675–682. [CrossRef]
25. Stadler, W.; Hofmann, D.M.; Alt, H.C.; Muschik, T.; Meyer, B.K.; Weigel, E.; Müller-Vogt, G.; Salk, M.; Rupp, E.; Benz, K.W. Optical investigations of defects in $Cd_{1-x}Zn_xTe$. *Phys. Rev. B* **1995**, *51*, 10619. [CrossRef]
26. Kuok, M.H.; Tan, L.S.; Shen, Z.X.; Huan, C.H.; Mok, K.F. A Raman study of $RbSnBr_3$. *Solid State Commun.* **1996**, *97*, 497–501. [CrossRef]

coatings

MDPI

Article

The Construction of an α-F_2O_3/Tubular g-C_3N_4 Z-Scheme Heterojunction Catalyst for the Efficient Photocatalytic Degradation of Tetracycline

Feng Xu [1], Kai Zhang [1], Kun Li [1], Hao Ju [2], Qian Xue [1], Xueqiang Qi [1,*] and Jinxia Jiang [3,*]

[1] School of Chemistry and Chemical Engineering, Chongqing University of Technology, Chongqing 400054, China; zephyr@stu.cqut.edu.cn (F.X.); zhangkai@stu.cqut.edu.cn (K.Z.); likun@stu.cqut.edu.cn (K.L.); xueq@stu.cqut.edu.cn (Q.X.)
[2] New Energy Technology of CAERI Co., Ltd., Lian Zhu Road, Guojiatuo Street, Lianjiang New District, Chongqing 400039, China; juhao@caeri.com.cn
[3] Chongqing Medical and Pharmaceutical College, Chongqing 400020, China
* Correspondence: xqqi@cqut.edu.cn (X.Q.); jiang106@cqu.edu.cn (J.J.)

Abstract: Morphological engineering and semiconductor coupling show significant potential to increase the photocatalytic performance of graphite carbon nitride (g-C_3N_4). In this work, a unique Z-scheme heterojunction photocatalyst composed of tubular g-C_3N_4 (TCN) and α-F_2O_3 was successfully synthesized. Combining the experimental results and characterization, we extensively investigated the charge transfer mechanism of the α-F_2O_3/tubular g-C_3N_4 (FO-TCN) heterojunctions and processes in the photocatalytic degradation of tetracycline (TC). The tubular morphology provided a larger specific surface area, enhancing the light absorption area and thus improving the exposure of the active sites. Not only was the light absorption range expanded through the coupling with α-F_2O_3, but the charge transfer properties of the sample were also strengthened. The synergism between photocatalysis and the Fenton reaction enhanced the photocatalytic performance of the FO-TCN. Due to the previously mentioned beneficial factors, the performance of the FO-TCN photocatalyst was significantly increased; its reaction rate k value in the degradation of TC (0.0482 min^{-1}) was 4.05 times faster than that of single g-C_3N_4 and it exhibited the best photocatalytic performance (95.02%) for the degradation of TC in 60 min, with an enhancement of 38.41%. Quenching experiments showed that h^+ and $\cdot O_2^-$ were the major active substances in the photocatalytic degradation process.

Keywords: tubular g-C_3N_4; α-F_2O_3; Z-scheme heterojunction; degradation of tetracycline

1. Introduction

In recent years, with industrial and urban development, a wide range of organic compounds have been introduced into the water environment, posing a serious threat to the ecosystem and to humans. Additionally, contamination with antibiotics has received extensive attention due to its threats to the environment [1–4]. Tetracycline (TC) is a very common antibiotic used in humans and animals and has a long history of use in the medical and animal feed industries. Typically, it is difficult for TC to be completely metabolized in the body due to its complex molecular structure and high stability, which leads to its widespread occurrence in the environment, inevitably resulting in environmental pollution [5,6]. Thus, there is a need for an advanced and environmentally friendly technology for the degradation of TC. In this context, semiconductor photocatalytic technology has developed into the most modern and environmentally friendly technology for water treatment [7–9].

To date, the use of semiconductor photocatalysts has attracted great interest worldwide in the field of the degradation of antibiotics, due to their eco-friendliness and cost-effectiveness [10,11]. In previous studies, the practical applications of conventional photocatalysts (ZnO and CdS) were limited due to their poor solar energy conversion performance and large band gap, especially the fast recombination rate of photogenerated e^-/h^+ [12,13]. Recently, graphite carbon nitride (g-C_3N_4) has gained widespread interest for its low production cost, ease of preparation, and suitable band gap [14,15]. However, the application of g-C_3N_4 is limited by its disadvantages, which include a volume structure with a small specific surface area, insufficient light absorption, and a low photogenerated carrier transfer capacity [16]. Hence, a number of strategies, including morphological adjustment, elemental doping, and combination with semiconductor materials, have been proposed to overcome these inherent disadvantages. Normally, photocatalytic processes take place at the surface of the photocatalyst, and the particular morphology of modified g-C_3N_4 exhibits excellent photocatalytic performance compared to its bulk counterparts [17–19]. Thus far, g-C_3N_4 photocatalysts in various nanostructured forms (nanosheet, nanotube, flower-like, microsphere, etc.) have been prepared successfully [20–24]. Wang et al. prepared novel, spongy g-C_3N_4 with a porous structure via protonated melamine annealing under an N_2/H_2 atmosphere, showing thinner nanosheets and a larger surface area [25]. Yang et al. prepared flower-like P-doped graphitic carbon nitride (g-C_3N_4) with a large surface area and high porosity via a facile thermal copolymerization method [26]. The larger surface area can provide more active sites directly and is conducive to shortening the diffusion distance of photogenerated carriers as well [27]. Due to the latest experimental research, g-C_3N_4 nanotubes have attracted great interest for their particular structure, which gives them excellent surface photocatalytic properties, promising electron transfer rates, and excellent optical qualities. However, liquid-phase reactions and template-assisted processes have generally been used to synthesize the previously studied g-C_3N_4 nanotubes [28]; these synthesis methods are inherently uncontrollable and difficult to perform because of the high-pressure conditions and the complex template residues required. In recent studies, a new, simple, and highly desirable method for the mass synthesis of tubular g-C_3N_4 was developed [29].

Currently, α-Fe_2O_3 is widely applied in photocatalytic composite material systems because of its non-toxicity, affordability, and property of absorbing visible light [30–35]. However, its widespread application as a highly efficient photocatalyst is restricted by the fast recombination of the photogenerated e^-/h+ pairs. Meanwhile, α-Fe_2O_3 presents a band gap in which the CB and VB edge potentials are close to 0.49 eV and 2.29 eV, respectively, which are lower than the corresponding values for g-C_3N_4 [36]. Theoretically, g-C_3N_4 could fit comfortably with Fe_2O_3, forming II-scheme (or Z-scheme) heterojunctions at their interface [37], and this allows for the energy levels to be staggered and the photogenerated carriers to be transferred [38]. g-C_3N_4 is combined with α-Fe_2O_3 to produce a α-Fe_2O_3/g-C_3N_4 heterojunction photocatalyst. In this situation, it plays an important role not only as a catalytic component but also as a viable and outstanding iron-based photo-Fenton catalyst [39]. The principle of the photo-Fenton reaction is that H_2O_2, in the presence of the catalyst Fe^{2+}, generates strong oxidizing power in the form of -OH and triggers the production of additional reactive oxygen species to achieve the degradation of organic matter. The reaction system can convert complexes such as $Fe(OH)^{2+}$ into Fe^{2+} under UV light irradiation so that Fe^{3+}/Fe^{2+} maintains a good cycle, which, in turn, accelerates the production of ·OH through H_2O_2 [40–42]. Thus, it is a feasible solution to construct an α-Fe_2O_3/g-C_3N_4 photocatalysis–Fenton system.

Herein, g-C_3N_4 is adopted as a supporting system for α-Fe_2O_3 based on its ease of modification and superior photocatalytic properties. In the present study, the structure and properties of pristine g-C_3N_4 are modified through morphology engineering, and FO-TCN samples are prepared. The increased specific surface area of the TCN offers more sites to load Fe_2O_3, and the loaded α-Fe_2O_3 has the benefits of the Fenton and the photocatalytic reactions. The scanning electron microscopy (SEM), transmission electron microscopy (TEM), X-ray diffraction (XRD), nitrogen adsorption/desorption, Fourier-transform infrared (FT-IR), X-ray photoelectron spectroscopy (XPS), and ultraviolet–visible diffuse reflection spectroscopy (UV–vis DRS) techniques were used for the characterization of the photocatalysts with FO-TCN. In terms of photocatalytic degradation activity, compared with another sample, the synthesized FO-TCN shows extraordinary photocatalytic performance for the degradation of TC. Furthermore, to investigate the active species' photocatalytic performance, radical trapping experiments were carried out. TBA, EDTA-2Na, L-tryptophan, and ASC were used as quenchers for ($\cdot OH$), (h^+), $^1O_2$, and ($\cdot O_2^-$), respectively. The results show that both h^+ and $\cdot O_2^-$ play important roles in the degradation of TC.

2. Experimental

2.1. Preparation of Photocatalysts

2.1.1. Synthesis of CN and FO-CN

Melamine (6 g) was heated in a muffle furnace to 550 °C at 2.3 °C·min^{-1} and calcined for 4 h. The product was then allowed to cool down to room temperature naturally. After obtaining the CN, 0.25 g iron nitrate and 0.3 g ammonium bicarbonate were added to 100 mL ethanol, and the mixture was then ultrasonicated for 30 min at 25 °C and stirred for 4 h. Finally, the samples were calcined using the same procedures described above.

2.1.2. Synthesis of Tubular TCN and FO-TCN

The TCN was synthesized through a precursor-reforming strategy [43]. Firstly, urea (8 g) and melamine (6 g) were placed in 40 mL deionized water, respectively. Then, the melamine and urea solutions were mixed together. Subsequently, the mixture was placed in a 100 mL autoclave lined with Teflon and maintained at 180 °C for 1 d. In the process of hydrothermal treatment, melamine can partly form cyanuric acid at 180 °C. The regular precursors of TCN were eventually produced by the self-assembly of the molecules between melamine and cyanuric acid. The precipitate was washed with 50 mL ethanol and then 50 mL deionized water three times. The precursor was then ground into a powder and calcined at 550 °C to form TCN. Based on the previous synthesis process, FO-TCN was synthesized as follows. After obtaining the precursors, a white sample, 0.25 g iron nitrate with 0.30 g ammonium bicarbonate, was added to 100 mL ethanol; then, the mixture was ultrasonically treated for 30 min and stirred for 4 h. Finally, it was heated following the same procedures as for TCN synthesis to obtain the FO-TCN sample.

2.2. Photocatalyst Characterization

XRD (XRD-7000, Shimadzu, Kyoto, Japan) was used to characterize the as-prepared powder catalysts. FI-TR was performed (Shimadzu, Japan, IRTracer-100) over the range of 400–4000 nm with the KBr pellet method. The morphologies of the as-prepared samples were tested by SEM (Phenom Pharos G1, Eindhoven, The Netherlands). The specific surface area, pore size and pore volume of the samples were obtained from nitrogen adsorption/desorption isotherms on a (KuBo-X1000, Bjbuilder, Beijing, China) machine. XPS (Thermo Escalab 250-Xi, Thermo, Wilmington, NC, USA) was used to analyze the surface composition and chemical state distribution of the samples, and C 1s was typically used as a reference for calibration. The light absorption of the as-prepared samples was analyzed with a UV–vis spectrophotometer (UV-2700i, Shimadzu, Suzhou, China) with the wavelength range of 200–800 nm.

2.3. Photocatalytic Performance Evaluation

The index used to evaluate the photocatalytic performance of the samples was the degradation efficiency of TC. A 300 W xenon lamp (Perfectlight, PLS-SXE300–/UV, Beijing, China) was used to simulate light energy. Typically, 20 mg photocatalyst was added to 100 mL TC (20 mg/L) solution and stirred for 60 min under dark conditions to reach adsorption/desorption equilibrium. Next, 50 μL H_2O_2 solution (30%) was added to the system. Then, the light energy was introduced into the system, and 3 mL of the suspension was taken at 5 min intervals. The suspension was then filtered by a 0.22 μm Millipore filter and finally evaluated using a spectrophotometer (Techcomp TU-1901, Shanghai, China) to test the concentration of TC at a wavelength of 357 nm. The formula for the calculation of the photocatalytic efficiency of the degradation of TC (η, %) is shown in Equation (1):

$$\eta = \frac{C_0 - C_t}{C_0} \times 100\% \quad (1)$$

where C_0 is the initial concentration of TC, while C_t is the concentration of TC after a degradation time of *t* minutes.

3. Results and Discussion

3.1. Physical and Chemical Properties

Characterization of the crystal structure and chemical molecular structure is necessary for the understanding of photocatalysts. XRD and FT-IR characterization were firstly performed. As shown in Figure 1a, the as-prepared samples had distinct diffraction peaks at 13.0° and 27.7°, corresponding to the respective (100) and (002) crystal planes of CN, which can be attributed to an in-plane arrangement of ordered tri-s-triazine units and the interlayer stacking of the conjugated aromatic system, respectively [44]. The characteristic peak of TCN at the 27.7° location was apparently weaker compared to the CN sample, which was attributed to the size effect originating from the ultra-thin tubular structure, thus causing a reduction in the stacking structure in the layer [45]. It appeared that the XRD spectra of both FO-CN and FO-TCN had two groups of peaks matching Fe_2O_3 and $g\text{-}C_3N_4$, respectively. The peaks at 24.2°, 33.2°, 35.8°, 41.0°, 49.6°, 54.2°, 57.7°, 62.5° and 64.1° of $\alpha\text{-}Fe_2O_3$ could be assigned to the (012), (104), (110), (113), (024), (116), (018), (214) and (300) crystal planes ($\alpha\text{-}Fe_2O_3$ JCPDS. No.33-0664). Notably, the characteristic peak at 27.7° in FO-TCN was slightly shifted because the Fe_2O_3 modulated the layer spacing, reflecting the fact that Fe_2O_3 was not confined to the shallow surface but had strong interfacial interactions with $g\text{-}C_3N_4$. After loading with the $\alpha\text{-}Fe_2O_3$, the sample displayed the corresponding characteristic peaks, while the diffraction peaks of $g\text{-}C_3N_4$ in the samples decreased, indicating that FO-TCN heterostructures had been successfully fabricated.

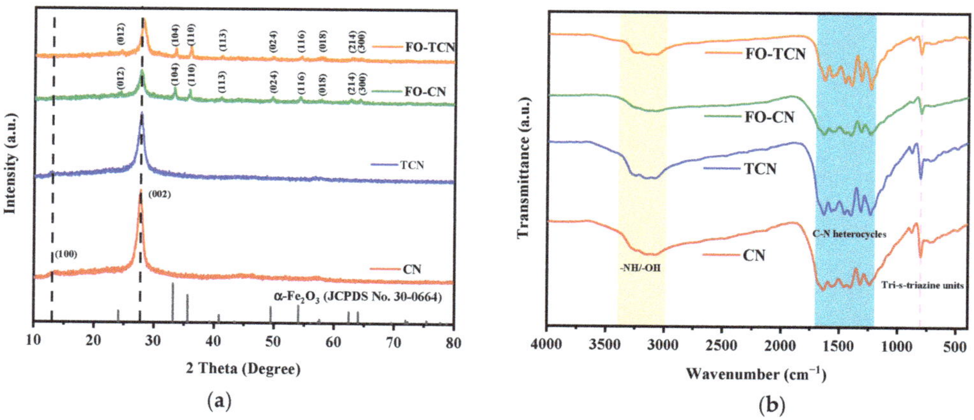

Figure 1. (**a**) XRD patterns and (**b**) FT-IR spectra of CN, TCN, FO-CN and FO-TCN, respectively.

To better illustrate the chemical molecular structure, the FT-IR spectra of the as-prepared samples are shown in Figure 1b. There were a number of main vibrational bands at 3000–3500 cm^{-1} corresponding to -NH and -OH stretching vibrations. The strong absorption peaks from 1200 to 1700 cm^{-1} could be attributed to the stretching vibration of the C-N heterocycles. Moreover, the sturdy absorption peak at 810 cm^{-1} was ascribed to the characteristic stretching vibration of tri-s-triazine units [46]. The typical absorption peaks of $g\text{-}C_3N_4$ were present in all the as-prepared samples. However, the corresponding peak position of the Fe-O bond in the composite was not obvious in the range of 450–1000 cm^{-1}. This could be attributed to the low loading of α-Fe_2O_3, which agreed with previously reported research [47,48]. A small shift in the peaks indicated that there were some interfacial interactions among α-Fe_2O_3 and g-C_3N_4 in the FO-TCN composite sample, which helped to form heterojunctions.

The morphologies and microstructures of CN, TCN, FO-CN and FO-TCN were examined by SEM. As shown in Figure 2a, CN presented an aggregated and chunky structure, while TCN formed an obvious tubular morphology (Figure 2b). Thus, the specific surface area can be increased and more active sites can be provided with the tubular morphology. It can be seen that α-F_2O_3 particles are distributed as aggregations on the surface of g-C_3N_4 in Figure 2c. Most importantly, it was clear that the FO-TCN retained the tubular structure when α-F_2O_3 was introduced to the TCN system. The close binding of α-F_2O_3 particles to the tubular surface of g-C_3N_4 can be observed in Figure 2d, which confirms the formation of a contact interface between the α-F_2O_3 particles and tubular g-C_3N_4.

Figure 2. SEM images of (**a**) CN, (**b**) TCN, (**c**) FO-CN and (**d**) FO-TCN, respectively.

In addition, to better illustrate the prepared tubular morphologies of the samples, the nitrogen adsorption/desorption isotherms and the Barrett—Joyner–Halenda (BJH) pore size distribution curves are shown in Figure 3a,b. As shown in Figure 3a, CN, TCN, FO-CN and FO-TCN all showed the typical type IV isotherm curves that characterize mesoporous materials. The pore volume of the original CN was smaller because of its bulk structure, whereas the modified TCN had a larger pore volume. The Brunauer–Emmett–Teller (BET) specific surface areas (S_{BET}) of CN, TCN, FO-CN and FO-TCN were calculated to be 21.569 $m^2 \cdot g^{-1}$, 50.127 $m^2 \cdot g^{-1}$, 16.091 $m^2 \cdot g^{-1}$ and 92.053 $m^2 \cdot g^{-1}$, respectively. In summary, the g-C_3N_4 morphology in the present study greatly increased the specific surface area, thereby offering more anchor points for α-F_2O_3. Notably, the specific surface area of FO-TCN was higher than those of TCN and BCN, suggesting that the introduction of α-F_2O_3 increased the specific surface area of the composite and thus provided more active sites. This tubular structure resulted in a large specific surface area, providing more active sites for surface reactions and facilitating photogenerated electron transfer over shorter distances, thereby significantly enhancing the charge separation and transfer efficiency.

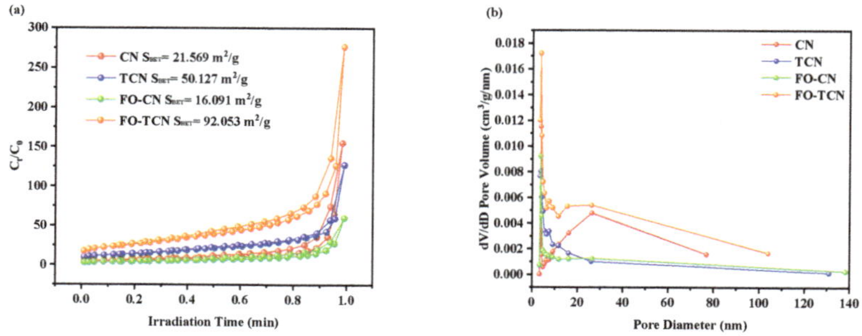

Figure 3. (**a**) Nitrogen adsorption/desorption isotherms and (**b**) pore size distribution curves.

XPS was applied to analyze the chemical state of the elements on the catalyst surfaces. As shown in Figure 4a, the elements of C, N, O and Fe could be determined in the XPS spectra of both FO-CN and FO-TCN. Then, the chemical states of FO-CN and FO-TCN were determined through high-resolution XPS spectroscopy. In the C 1s spectrum exhibited in Figure 4b, the two major peaks located at 284.8 eV and 288.3 eV belong to occasional graphitic carbon atoms (C-C) and sp^2 hybridized carbon atoms of the aromatic rings (N-C=N), respectively. In addition, a peak at 286.1 eV corresponding to the sp^3-coordinated carbon (C-N) can be determined. In Figure 4c, N 1s of FO-CN shows that the main peak at 398.6 eV was identified as the sp^2 hybridized N of C-N=C, the second one at 400.0 eV was ascribed to the N-C_3 group and the last one at 400.8 eV was attributed to the terminal C-N=C carbon atom, while the peak of FO-TCN was slightly shifted towards a larger binding energy value. The O 1s of both FO-CN and FO-TCN are shown in Figure 4d. FO-CN presented two peaks at 529.9 and 531.9 eV corresponding to lattice oxygen and surface hydroxyl groups, respectively, while the binding energy of the lattice oxygen of FO-TCN shows a 0.2 eV negative shift. In Figure 4e, the Fe 2p of FO-CN shows two distinct peaks at 711.1 and 724.9 eV, and two oscillating satellite peaks at 717.3 and 732.2 eV, indicating that the Fe ions in FO-CN have a +3 value, while the Fe $2P_{3/2}$ of FO-TCN shows a weak positive shift. UV–vis DRS was used to characterize the optical absorption properties of the materials. As shown in Figure 5a, both CN and TCN had absorption edges at around 450 nm, respectively, which agreed with the previously reported absorption edges. Moreover, the introduction of α-F_2O_3 caused a red shift in the absorption edges of FO-CN and FO-TCN. This may have been caused by the synergistic effect between α-F_2O_3 and g-C_3N_4, which can expand the light absorption range of composite materials. Thus, the ability of the material to absorb visible light was enhanced through the introduction of Fe_2O_3, which is beneficial to the generation of charge carriers. Furthermore, the Kubelka–Munk function was used through the conversion of UV–vis DRS data to obtain the band gap energy of the photocatalyst, and the band gaps of CN, TCN, FO-CN and FO-TCN were 2.69 eV, 2.63 eV, 2.29 eV and 2.40 eV, respectively (Figure 5b). As shown in Figure S1, every average relative deviation of the band gap in the CN, TCN, FO-CN and FO-TCN systems was less than 2%. The band gap of the prepared composite was decreased after the introduction of α-F_2O_3 with a band gap of 2.01 eV (Figure 5c) into the system, which was beneficial to the photocatalyst for the further absorption of sunlight and the improvement of the photocatalytic performance.

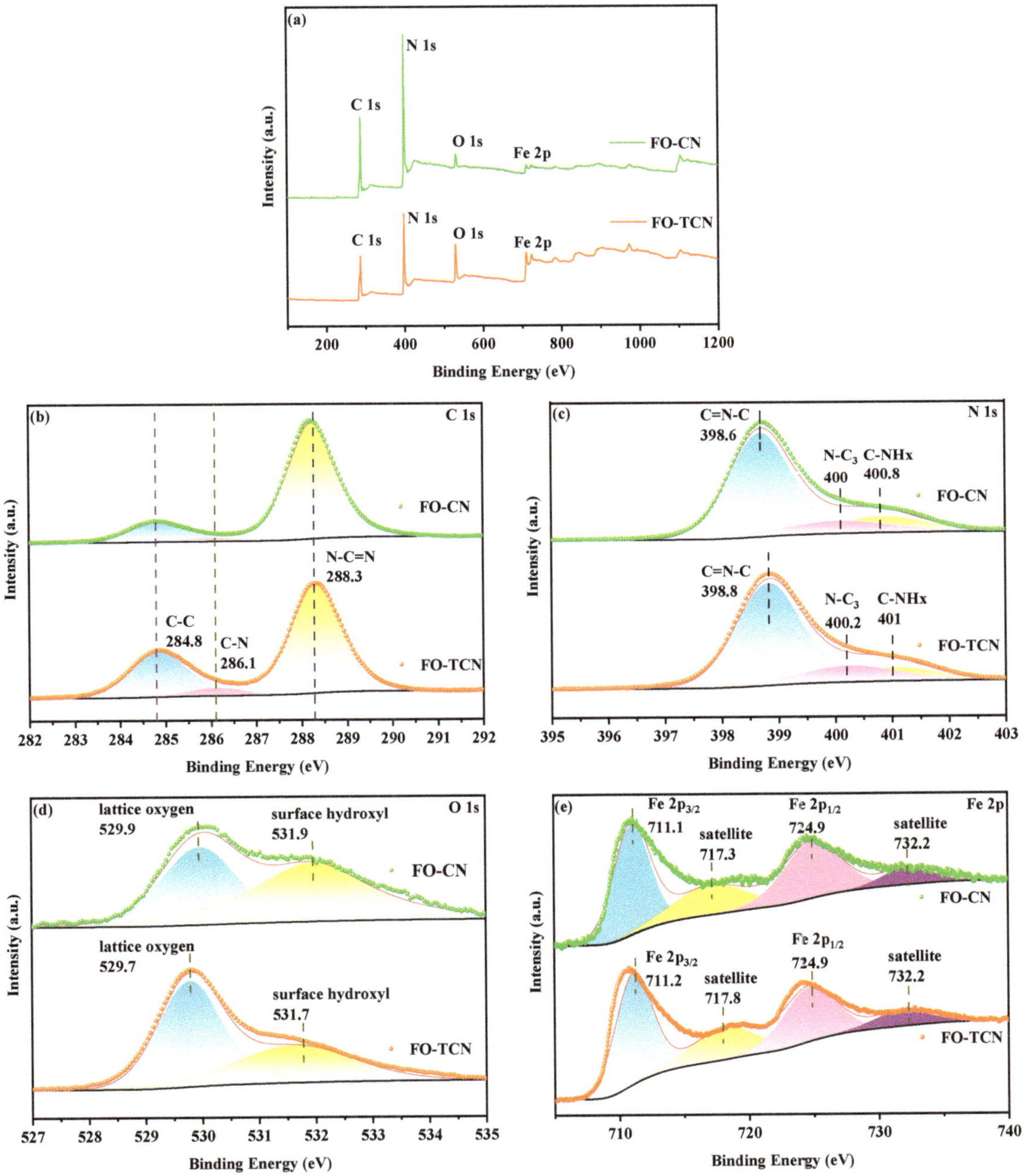

Figure 4. XPS spectra of FO-CN and FO-TCN: (**a**) survey, (**b**) C 1s, (**c**) N 1s, (**d**) O 1s and (**e**) Fe 2p.

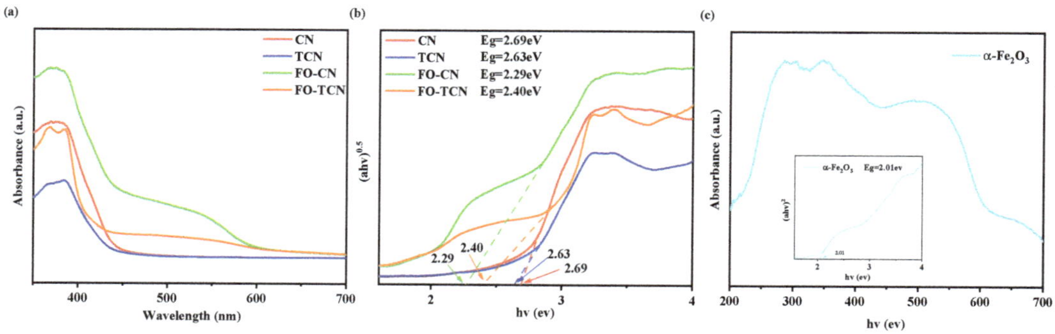

Figure 5. (**a**) UV–vis DRS of samples, (**b**) corresponding band gap energy diagram of CN, TCN, FO-CN, FO-TCN and (**c**) UV–vis DRS of pristine α-Fe_2O_3.

The conduction band (CB) and valence band (VB) edge potentials of the photocatalysts were calculated according to the empirical Equations (2) and (3):

$$E_{CB} = X - E_e - 0.5E_g \tag{2}$$

$$E_{VB} = E_{CB} + E_g \tag{3}$$

X is the absolute potential of the semiconductor, E_e is the electron free energy relative to the hydrogen potential level (4.5 eV) and E_g is the band gap of the semiconductor, which was obtained through UV–vis DRS in our experiment. Thus, it can be deduced that the CB and VB edge potentials of TCN were −1.02 eV and +1.61 eV (vs. NHE), while the CB and VB edge potentials of α-F_2O_3 were +0.38 eV and +2.39 eV (vs. NHE), respectively.

3.2. Photocatalytic Activity and Mechanisms

In order to investigate the photocatalytic ability of the samples, the photocatalytic degradation of TC was performed. The adsorption/desorption equilibrium of TC with the photocatalyst was reached under the conditions of 60 min dark treatment before light irradiation [49]. Figure 6a shows the variation in the TC concentration (C_t/C_0) with light irradiation time over the as-prepared photocatalysts. Previous studies have shown that TC is a stable compound that is poorly metabolized in the body and cannot be decomposed in the absence of photocatalysts under light irradiation [7]. The different photocatalysts showed different photodegradation efficiency in 60 min, which was in the order of CN (56.61%) < TCN (78.77%) < FO-CN (82.28%) < FO-TCN (95.02%). In the 60 min photocatalytic process, the degradation efficiency of CN was the lowest (56.61%). The degradation efficiency of TCN was 22.16% higher than that of CN, which showed that the unique tubular morphology of TCN increased the photocatalytic capacity of TCN. As expected, all composites displayed superior photocatalytic decomposition efficiency compared to both pristine CN and TCN, which proves that the introduction of α-F_2O_3 to g-C_3N_4 is valuable. FO-TCN exhibited the best photocatalytic performance (95.02%) for the degradation of TC in 60 min, with an enhancement of 38.41% and 16.25% compared to CN and TCN. The apparent rate constant (k) can be calculated with Equation (4):

$$k = \frac{\ln(C_t/C_0)}{t} \tag{4}$$

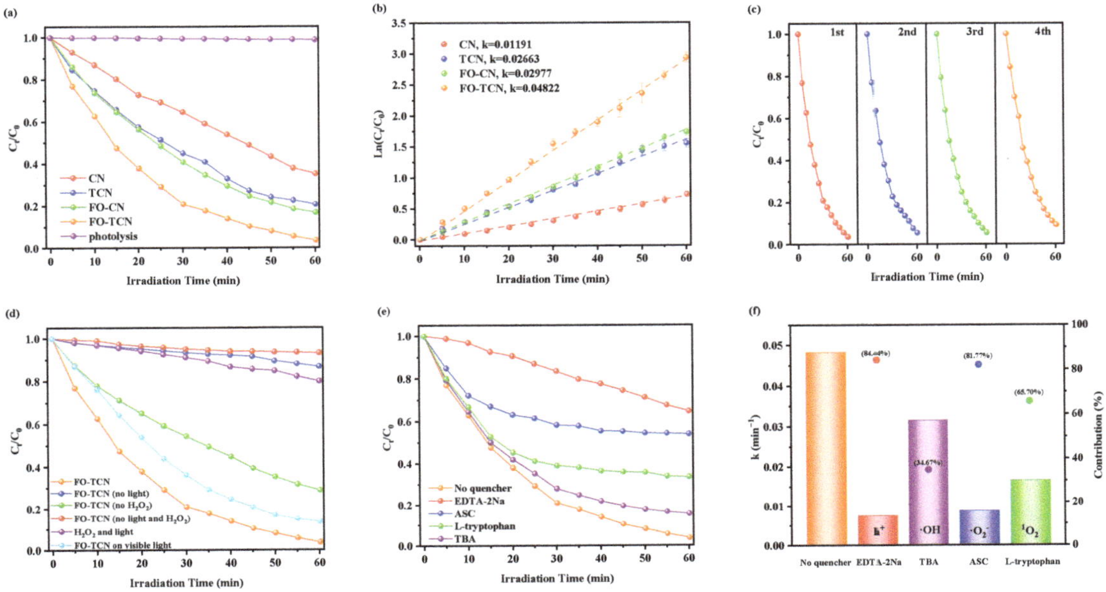

Figure 6. (**a**) Photocatalytic degradation curves of TC over different samples, (**b**) kinetic curves and the corresponding apparent rate constants, (**c**) the cycle experiment for the degradation of TC in FO-TCN system, (**d**) photocatalytic degradation experiments with different treatment conditions, (**e**) degradation of TC with different quenching agents, (**f**) corresponding kinetic constants and the relative contributions of different quenchers.

As can be seen from the kinetic curves in Figures 6b and S2, the average value of k followed the sequence of CN (0.0119 min^{-1}) < TCN (0.0266 min^{-1}) < FO-CN (0.0298 min^{-1}) < FO-TCN (0.0482 min^{-1}), and every average relative deviation in k in the CN, TCN, FO-CN and FO-TCN systems was less than 1%. FO-TCN indicated the highest k value (0.0482 min^{-1}), which was 4.05 and 1.81 times higher than that of CN and TCN, respectively. The above experimental results show that FO-TCN had the strongest ability for TC degradation. The rate of TC degradation and the value of the rate constant k of FO-TCN were clearly higher than those of the other samples. This result proves that the regulation of the morphology is an important factor in the improvement of the photocatalytic performance. The tubular structure could not only improve the light absorption area but also expand the contact area with TC contaminants. Meanwhile, the addition of α-F_2O_3 to g-C_3N_4 was beneficial because the photocatalytic performance of FO-TCN was enhanced, which was attributed to the synergistic effect of enhancing the light response and improving the efficiency of photogenerated carrier separation. As is well known, an important measurement of the practical application of a catalyst is its reusability. Thus, we carried out a cycling experiment. As shown in Figure 6c, FO-TCN achieved degradation efficiency of 90.91% after four cycles, reflecting the good reusability of the sample.

As shown in Figure 6d, the effects of the light energy, H_2O_2 and photocatalyst on TC degradation were studied by conducting degradation experiments in different experimental situations. The experimental results showed that only the FO-TCN photocatalyst was not able to degrade TC without light energy. Meanwhile, in the absence of light energy, the photocatalytic degradation of TC by FO-TCN supported by H_2O_2 alone in the Fenton system was low, proving that the efficiency of the Fenton reaction was not optimal. In addition, it showed that H_2O_2 could slightly decompose TC under light energy without the addition of a photocatalyst. This was due to the oxidation of H_2O_2 itself to generate hydroxyl radicals (·OH) under light energy. Furthermore, the degradation rate of TC over FO-TCN decreased clearly in the absence of H_2O_2, which demonstrated that TC decomposition was accelerated by the photo-Fenton reaction. In addition, the degradation of TC in the range of visible light was studied comparatively. The slightly decreased efficiency (87.3%) of TC degradation was observed under visible light irradiation. The reason lies in the fact that not only is the full spectrum more energetic, but the g-C_3N_4 has some light absorption below 420 nm.

In order to study the contributions of the different oxidative species generated to the degradation of TC in the FO-TCN system, the addition of quenchers was used to perform active species trapping tests. Herein, ethylene diamine tetra-acetic acid (EDTA-2Na), L-ascorbic acid (ASC), L-tryptophan and tert-butyl alcohol (TBA) were used to quench h^+, $\cdot O_2^-$, $^1O_2$ and $\cdot OH$, respectively [50]. The photocatalytic degradation efficiencies of TC over the FO-TCN photocatalyst in the presence of these four quenchers are shown in Figure 6e,f. When ASC and EDTA-2Na were introduced into the photocatalytic system, the degradation efficiency of TC dropped significantly from 95.02% to 46.15% and to 35.38%, respectively. Furthermore, the degradation efficiency of TC was also decreased from 95.02% to 66.56% in the presence of L-tryptophan. These experiments indicated that both $\cdot O_2^-$ and h^+ had a very important effect on the photocatalytic degradation of TC for the FO-TCN photocatalyst, while $^1O_2$ also played a certain of role in the process of TC degradation. As shown in Figure 6f, the corresponding apparent kinetic constants were 0.0482, 0.0750, 0.0879, 0.0165, 0.0315 min^{-1}, respectively. Furthermore, the relative contribution (R) of different active species to the degradation of TC was calculated through Equations (5)–(8) [44]:

$$R_{\cdot O_2^-} = \frac{k_{\cdot O_2^-}}{k} \approx \frac{(k - k_{ASC})}{k} \tag{5}$$

$$R_{\cdot OH} = \frac{k_{\cdot OH}}{k} \approx \frac{(k - k_{TBA})}{k} \tag{6}$$

$$R_{h^+} = \frac{k_{h^+}}{k} \approx \frac{(k - k_{EDTA\text{-}2Na})}{k} \tag{7}$$

$$R_{^1O_2} = \frac{k_{^1O_2}}{k} \approx \frac{\left(k - k_{L\text{-}tryptophan}\right)}{k} \tag{8}$$

The corresponding contribution rates of h^+, $\cdot O_2^-$, $^1O_2$ and $\cdot OH$ to the TC removal were calculated to be 84.44%, 81.17%, 65.70% and 34.67%, respectively. These results show that both h^+ and $\cdot O_2^-$ played the dominant role in the FO-TCN photocatalytic system for the degradation of TC, while $^1O_2$ played a subordinate role that also cannot be ignored. The strong electron coupling effect was responsible for the abundance of active species discovered in the FO-TCN heterojunction, which stimulated the photogenerated charge kinetics and provided a large number of reactive centers to accelerate the degradation reaction.

The possible photocatalytic reaction mechanism of the FO-TCN heterojunction is illustrated in Figure 7 according to Equations (9)–(17). Based on the data from the trapping experiment, -O_2^- was a crucial active substance in the FO-TCN system and had a significant influence on the photocatalytic degradation of TC. According to the previous results derived from Figure 5, the CB and VB edge potentials of TCN were −1.02 eV and +1.61 eV (vs. NHE), while the CB and VB edge potentials of α-F_2O_3 were +0.38 eV and +2.39 eV (vs. NHE), so they had a staggered energy band arrangement.

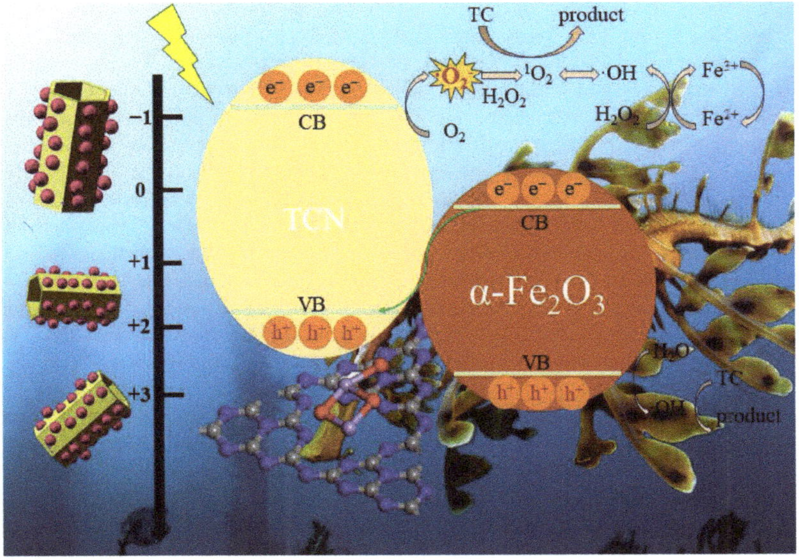

Figure 7. The proposed mechanism of TC degradation by the FO-TCN system.

In the typical type II heterojunction, FO-TCN is not able to form $\cdot O_2^-$, due to the fact that the redox potential of $O_2/\cdot O_2^-$ is lower compared to the CB of α-F_2O_3, and the electrons pooled on the CB of α-F_2O_3 do not have sufficient power to produce the $\cdot O_2^-$ free radical, apparently contradicting the conclusions of the free radical scavenging experiments [51].

$$\text{FO-TCN} + hv \rightarrow \text{FO-TCN}\left(e^- + h^+\right) \tag{9}$$

$$e^- + O_2 \rightarrow \cdot O_2^- \tag{10}$$

$$\cdot O_2^- + H_2O_2 \rightarrow {}^1O_2 \tag{11}$$

$$H_2O_2 + e^- \rightarrow \cdot OH + OH^- \tag{12}$$

$$OH^- + h^+ \rightarrow \cdot OH \tag{13}$$

$$\equiv Fe^{3+} + e^- \rightarrow Fe^{2+} \tag{14}$$

$$\equiv Fe^{2+} + H_2O_2 \rightarrow Fe^{2+} + \cdot OH + OH^- \tag{15}$$

$$\equiv Fe^{3+} + H_2O_2 \rightarrow Fe^{2+} + \cdot OOH + h^+ \tag{16}$$

$$\left(h^+, \cdot O_2^-, {}^1O_2, \cdot OH\right) + TC \rightarrow \text{Products} \tag{17}$$

Meanwhile, the presence of $\cdot O_2^-$ was demonstrated in the results of the radical scavenging experiment. Therefore, it is easy to conclude that the carrier transfer pathway between the FO-TCN heterojunction synthesized was in accordance with the charge transfer mechanism of the Z-scheme, which fully maintained the redox performance of the catalyst. Both g-C_3N_4 and α-F_2O_3 can be excited to generate photo-induced e^- and h^+

under light radiation. At the interface between two components, the e^- in the CB of α-F_2O_3 can move to the VB of g-C_3N_4 via an internal electric field, resulting in strong electrostatic attraction between e^- in the CB of α-F_2O_3 and h^+ in the VB of g-C_3N_4. This process reduces the recombination of e^- and h^+ pairs by increasing the separation of the charge carriers e^- in the CB of g-C_3N_4 and h^+ in the VB of α-F_2O_3. More importantly, the h^+ itself is a highly oxidizing agent and can directly oxidize pollutants to achieve degradation. In addition, compared to the standard redox potential of O_2/O_2^-, the electrons in the CB of g-C_3N_4 can be trapped with O_2 to produce $\cdot O_2^-$, since the CB potential of g-C_3N_4 is more negative. Then, the $\cdot O_2^-$ radical is combined with H_2O_2 to form ($^1O_2$), which is compatible with the Fenton reaction mechanism. Finally, the portion of photogenerated electrons of α-F_2O_3 could enhance the photo-Fenton catalytic process through participation in the Fe^{3+}/Fe^{2+} cycle process, and they can activate H_2O_2 to produce $\cdot OH$, while forming a typical Fenton system. The photo-Fenton reaction has higher catalytic activity in the degradation of TC compared to the purely photocatalytic reaction. In addition, the continuous conversion of Fe^{3+}/Fe^{2+} could be maintained by the introduction of H_2O_2, which is the critical factor to trigger the photo-Fenton reaction. Only trace amounts of H_2O_2 are required to promote the photocatalytic performance of FO-TCN, thus eliminating the cost of adding H_2O_2. Furthermore, since the VB of Fe_2O_3 has a more positive potential than $\cdot OH/OH^-$ and $\cdot OH/H_2O$, the vacancies accumulated on the VB may oxidize OH^- and H_2O to generate $\cdot OH$.

4. Conclusions

A Z-scheme heterojunction of FO-TCN, which showed outstanding catalytic activity in the photocatalytic degradation of TC, was successfully synthesized through a facile precursor-reforming strategy. According to the sample characterization, the tubular morphology offered samples with a large specific surface area, thus providing a larger contact surface for contaminants and increasing the number of active sites. In addition, the particular tubular morphology reduced the transfer pathway distance of photogenerated carriers and enhanced the interaction with the reactant. Under light irradiation, FO-TCN achieved 95.02% efficiency for the degradation of TC in 60 min, which was clearly higher than that of pristine g-C_3N_4 (56.61%) under the same conditions, and it obtained the highest k value (0.0482 min^{-1}), which was 4.05 times higher than that of pristine g-C_3N_4. The FO-TCN heterojunction not only shortened the time and path of charge transfer, but also provided sufficient active reaction sites. Furthermore, the major active species $\cdot O_2^-$ and h^+ were the main active species during the photocatalytic degradation of TC, while the rapid Fe^{3+}/Fe^{2+} conversion caused by H_2O_2 facilitated the formation of $\cdot OH$, thereby increasing the photocatalytic degradation effect at the same time. Consequently, the present research is very significant in promoting the widespread application of g-C_3N_4-based heterojunctions in photocatalytic purification treatment, and the technology for the combination of magnetic ferrites with g-C_3N_4 could also lead to industrial recycling applications in the future. Moreover, the FO-TCN photocatalysts synthesized in this study have excellent properties and practical uses, such as the removal of antibiotics and pharmaceuticals from wastewater.

Supplementary Materials: The following supporting information can be downloaded at: https://www.mdpi.com/article/10.3390/coatings13111909/s1, Figure S1. The band gap energy diagram of (a) CN, (b) TCN, (c) FO-CN, and (d) FO-TCN, respectively. Figure S2. Kinetic curve and the corresponding apparent rate constant of (a) CN, (b) TCN, (c) FO-CN, and (d) FO-TCN, respectively.

Author Contributions: Conceptualization, F.X. and X.Q.; methodology, F.X. and J.J.; formal analysis, K.Z., K.L., H.J. and Q.X.; data curation, F.X. and Q.X.; writing—original draft preparation, F.X. and X.Q.; writing—review and editing, F.X., X.Q. and J.J.; supervision, X.Q. All authors have read and agreed to the published version of the manuscript.

Funding: This work was supported by the Natural Science Foundation of Chongqing (2023NSCQ-MSX1669).

Institutional Review Board Statement: Not applicable.

Informed Consent Statement: Not applicable.

Data Availability Statement: Data are contained within the article.

Acknowledgments: F.X. acknowledges the support of Chongqing University of Technology's Action Plan for Quality Development of Postgraduate Education (gzlcx20233170).

Conflicts of Interest: The authors declare no conflict of interest.

References

1. Yang, Y.; Li, X.; Zhou, C.; Xiong, W.; Zeng, G.; Huang, D.; Zhang, C.; Wang, W.; Song, B.; Tang, X.; et al. Recent advances in application of graphitic carbon nitride-based catalysts for degrading organic contaminants in water through advanced oxidation processes beyond photocatalysis: A critical review. *Water Res.* **2020**, *184*, 116200. [CrossRef] [PubMed]
2. Larsson, D.J.; Flach, C.F. Antibiotic resistance in the environment. *Nat. Rev. Microbiol.* **2022**, *20*, 257–269. [CrossRef] [PubMed]
3. Wang, J.; Zhuan, R. Degradation of antibiotics by advanced oxidation processes: An overview. *Sci. Total Environ.* **2020**, *701*, 135023. [CrossRef] [PubMed]
4. Zhang, C.; Lai, C.; Zeng, G.; Huang, D.; Yang, C.; Wang, Y.; Zhou, Y.; Cheng, M. Efficacy of carbonaceous nanocomposites for sorbing ionizable antibiotic sulfamethazine from aqueous solution. *Water Res.* **2016**, *95*, 103–112. [CrossRef] [PubMed]
5. Saadati, F.; Keramati, N.; Ghazi, M.M. Influence of parameters on the photocatalytic degradation of tetracycline in wastewater: A review. *Crit. Rev. Environ. Sci. Technol.* **2016**, *46*, 757–782. [CrossRef]
6. Jiang, H.; Wang, Q.; Chen, P.; Zheng, H.; Shi, J.; Shu, H.; Liu, Y. Photocatalytic degradation of tetracycline by using a regenerable (Bi)BiOBr/rGO composite. *J. Clean. Prod.* **2022**, *339*, 130771. [CrossRef]
7. Yang, Y.; Zeng, Z.; Zhang, C.; Huang, D.; Zeng, G.; Xiao, R.; Lai, C.; Zhou, C.; Guo, H.; Xue, W.; et al. Construction of iodine vacancy-rich BiOI/Ag@AgI Z-scheme heterojunction photocatalysts for visible-light-driven tetracycline degradation: Transformation pathways and mechanism insight. *Chem. Eng. J.* **2018**, *349*, 808–821. [CrossRef]
8. Calvete, M.J.F.; Piccirillo, G.; Vinagreiro, C.S.; Pereira, M.M. Hybrid materials for heterogeneous photocatalytic degradation of antibiotics. *Coord. Chem. Rev.* **2019**, *395*, 63–85. [CrossRef]
9. He, X.; Kai, T.; Ding, P. Heterojunction photocatalysts for degradation of the tetracycline antibiotic: A review. *Environ. Chem. Lett.* **2021**, *19*, 4563–4601. [CrossRef]
10. Baaloudj, O.; Assadi, I.; Nasrallah, N.; El Jery, A.; Khezami, L.; Assadi, A.A. Simultaneous removal of antibiotics and inactivation of antibiotic-resistant bacteria by photocatalysis: A review. *J. Water Process Eng.* **2021**, *42*, 102089. [CrossRef]
11. Li, D.; Shi, W. Recent developments in visible-light photocatalytic degradation of antibiotics. *Chin. J. Catal.* **2016**, *37*, 792–799. [CrossRef]
12. Rabell, G.O.; Cruz, M.R.A.; Juárez-Ramírez, I. Hydrogen production of ZnO and ZnO/Ag films by photocatalysis and photoelectrocatalysis. *Mater. Sci. Semicond. Process.* **2021**, *134*, 105985. [CrossRef]
13. Tang, X.; Xue, Q.; Qi, X.; Cheng, C.; Yang, M.; Yang, T.; Chen, F.; Qiu, F.; Quan, X. DFT and experimental study on visible-light driven photocatalysis of rare-earth-doped TiO_2. *Vacuum* **2022**, *200*, 110972. [CrossRef]
14. Guo, T.; Wang, K.; Zhang, G.; Wu, X. A novel α-Fe_2O_3@g-C_3N_4 catalyst: Synthesis derived from Fe-based MOF and its superior photo-Fenton performance. *Appl. Surf. Sci.* **2019**, *469*, 331–339. [CrossRef]
15. Papailias, I.; Todorova, N.; Giannakopoulou, T.; Ioannidis, N.; Dallas, P.; Dimotikali, D.; Trapalis, C. Novel torus shaped g-C_3N_4 photocatalysts. *Appl. Catal. B Environ.* **2020**, *26*, 118733. [CrossRef]
16. Liu, C.; Huang, H.; Cui, W.; Dong, F.; Zhang, Y. Band structure engineering and efficient charge transport in oxygen substituted g-C_3N_4 for superior photocatalytic hydrogen evolution. *Appl. Catal. B Environ.* **2018**, *230*, 115–124. [CrossRef]
17. Liu, C.; Huang, H.; Ye, L.; Yu, S.; Tian, N.; Du, X.; Zhang, T.; Zhang, Y. Intermediate-mediated strategy to horn-like hollow mesoporous ultrathin g-C_3N_4 tube with spatial anisotropic charge separation for superior photocatalytic H_2 evolution. *Nano Energy* **2017**, *41*, 738–748. [CrossRef]
18. Liang, Q.; Li, Z.; Yu, X.; Huang, Z.-H.; Kang, F.; Yang, Q.-H. Macroscopic 3D Porous Graphitic Carbon Nitride Monolith for Enhanced Photocatalytic Hydrogen Evolution. *Adv. Mater.* **2015**, *27*, 4634–4639. [CrossRef]
19. Nguyen, T.K.A.; Pham, T.-T.; Gendensuren, B.; Oh, E.-S.; Shin, E.W. Defect engineering of water-dispersible g-C_3N_4 photocatalysts by chemical oxidative etching of bulk g-C_3N_4 prepared in different calcination atmospheres. *J. Mater. Sci. Technol.* **2022**, *103*, 232–243. [CrossRef]
20. Wang, Y.; Li, L.; Wei, Y.; Xue, J.; Chen, H.; Ding, L.; Caro, J.; Wang, H. Water Transport with Ultralow Friction through Partially Exfoliated g-C_3 N_4 Nanosheet Membranes with Self-Supporting Spacers. *Angew. Chem. Int. Ed. Engl.* **2017**, *56*, 8974–8980. [CrossRef]
21. Guo, S.; Deng, Z.; Li, M.; Jiang, B.; Tian, C.; Pan, Q.; Fu, H. Phosphorus-Doped Carbon Nitride Tubes with a Layered Micro-nanostructure for Enhanced Visible-Light Photocatalytic Hydrogen Evolution. *Angew. Chem. Int. Ed. Engl.* **2016**, *55*, 1830–1834. [CrossRef] [PubMed]
22. Wang, H.; Lin, Q.; Yin, L.; Yang, Y.; Qiu, Y.; Lu, C.; Yang, H. Biomimetic Design of Hollow Flower-Like g-C_3N_4@PDA Organic Framework Nanospheres for Realizing an Efficient Photoreactivity. *Small* **2019**, *15*, e1900011. [CrossRef] [PubMed]

23. Si, Y.; Sun, Z.; Huang, L.; Chen, M.; Wu, L. A "ship-in-a-bottle" strategy to fabricate highly crystallized nanoporous graphiti C_3N_4 microspheres under pressurized conditions. *J. Mater. Chem. A* **2019**, *7*, 8952–8959. [CrossRef]
24. Patnaik, S.; Sahoo, D.P.; Parida, K. Recent advances in anion doped g-C_3N_4 photocatalysts: A review. *Carbon* **2021**, *172*, 682–71 [CrossRef]
25. Wang, C.; Han, T.; Xin, C.; Miao, H. Synthesizing the High Surface Area g-C_3N_4 for Greatly Enhanced Hydrogen Productio *Catalysts* **2021**, *11*, 832. [CrossRef]
26. Yang, H.; Zhou, Y.; Wang, Y.; Hu, S.; Wang, B.; Liao, Q.; Li, H.; Bao, J.; Ge, G.; Jia, S. Three-dimensional flower-like phosphorus doped g-C_3N_4 with a high surface area for visible-light photocatalytic hydrogen evolution. *J. Mater. Chem. A* **2018**, *6*, 16485–1649 [CrossRef]
27. Roškarič, M.; Zavašnik, J.; Zámbó, D.; Kotnik, T.; Kovačič, S.; Žerjav, G.; Pintar, A. Optimization Method Based on Simplex fo Surface Area Improved Photocatalytic Performance of g-C_3N_4. *ACS Catal.* **2023**, *13*, 13282–13300. [CrossRef]
28. Zhou, C.; Shi, R.; Shang, L.; Wu, L.-Z.; Tung, C.-H.; Zhang, T. Template-free large-scale synthesis of g-C_3N_4 microtubes fo enhanced visible light-driven photocatalytic H_2 production. *Nano Res.* **2018**, *11*, 3462–3468. [CrossRef]
29. Wang, W.; Zeng, Z.; Zeng, G.; Zhang, C.; Xiao, R.; Zhou, C.; Xiong, W.; Yang, Y.; Lei, L.; Liu, Y.; et al. Sulfur doped carbon quantum dots loaded hollow tubular g-C_3N_4 as novel photocatalyst for destruction of Escherichia coli and tetracycline degradation unde visible light. *Chem. Eng. J.* **2019**, *378*, 122132. [CrossRef]
30. Pham, V.V.; Truong, T.K.; Hai, L.V.; La, H.P.P.; Nguyen, H.T.; Lam, V.Q.; Tong, H.D.; Nguyen, T.Q.; Sabbah, A.; Chen, K.H.; et a S-Scheme α-Fe_2O_3/g-C_3N_4 Nanocomposites as Heterojunction Photocatalysts for Antibiotic Degradation. *ACS Appl. Nano Mate* **2022**, *5*, 4506–4514. [CrossRef]
31. Gan, P.; Zhang, Z.; Hu, Y.; Li, Y.; Ye, J.; Tong, M.; Liang, J. Insight into the role of Fe in the synergetic effect of persulfate/sulfite and Fe_2O_3@g-C_3N_4 for carbamazepine degradation. *Sci. Total Environ.* **2022**, *819*, 152787. [CrossRef] [PubMed]
32. Zhu, S.; Zou, X.; Zhou, Y.; Zeng, Y.; Long, Y.; Yuan, Z.; Wu, Q.; Li, M.; Wang, Y.; Xiang, B. Hydrothermal synthesis of graphene encapsulated 2D circular nanoplates of α-Fe_2O_3 towards enhanced electrochemical performance for supercapacitor. *J. Alloys Compd.* **2019**, *775*, 63–71. [CrossRef]
33. Xu, Q.-Q.; Huo, W.; Li, S.-S.; Fang, J.-H.; Li, L.; Zhang, B.-Y.; Zhang, F.; Zhang, Y.-X.; Li, S.-W. Crystal phase determined Fe active sites on Fe_2O_3 (γ- and α-Fe_2O_3) yolk-shell microspheres and their phase dependent electrocatalytic oxygen evolution reaction *Appl. Surf. Sci.* **2020**, *533*, 147368. [CrossRef]
34. Liu, S.; Wang, S.; Jiang, Y.; Zhao, Z.; Jiang, G.; Sun, Z. Synthesis of Fe_2O_3 loaded porous g-C_3N_4 photocatalyst for photocatalytic reduction of dinitrogen to ammonia. *Chem. Eng. J.* **2019**, *373*, 572–579. [CrossRef]
35. Zhu, H.; Chen, Z.; Hu, Y.; Gong, L.; Li, D.; Li, Z. A novel immobilized Z-scheme P_3HT/α-Fe_2O_3 photocatalyst array: Study on the excellent photocatalytic performance and photocatalytic mechanism. *J. Hazard. Mater.* **2020**, *389*, 122119. [CrossRef] [PubMed]
36. Ge, F.; Li, X.; Wu, M.; Ding, H.; Li, X. A type II heterojunction alpha-Fe_2O_3/g-C_3N_4 for the heterogeneous photo-Fenton degradation of phenol. *RSC Adv.* **2022**, *12*, 8300–8309. [CrossRef] [PubMed]
37. Geng, Y.; Chen, D.; Li, N.; Xu, Q.; Li, H.; He, J.; Lu, J. Z-Scheme 2D/2D α-Fe_2O_3/g-C_3N_4 heterojunction for photocatalytic oxidation of nitric oxide. *Appl. Catal. B Environ.* **2021**, *280*, 119409. [CrossRef]
38. Zhang, J.; Jiang, X.; Gong, M.; Zheng, Y. Z-scheme Fe_2O_3/g-C_3N_4 heterojunction with excellent photocatalytic property. *J Photonics Energy* **2020**, *10*, 023509. [CrossRef]
39. Wang, Y.; Song, H.; Chen, J.; Chai, S.; Shi, L.; Chen, C.; Wang, Y.; He, C. A novel solar photo-Fenton system with self-synthesizing H_2O_2: Enhanced photo-induced catalytic performances and mechanism insights. *Appl. Surf. Sci.* **2020**, *512*, 154650. [CrossRef]
40. Wang, W.; Zhao, W.; Huang, H.; Chen, R.; Shi, H. A 2D/2D S-scheme photo-Fenton catalyst based on ultrathin Bi2MoO_6 and Fe_2O_3 hexagonal nanosheets for efficient tetracycline degradation. *Catal. Sci. Technol.* **2021**, *11*, 2948–2956. [CrossRef]
41. Wang, W.; Zhao, W.; Zhang, H.; Dou, X.; Shi, H. 2D/2D step-scheme α-Fe_2O_3/Bi_2WO_6 photocatalyst with efficient charge transfer for enhanced photo-Fenton catalytic activity. *Chin. J. Catal.* **2021**, *42*, 97–106. [CrossRef]
42. Zhang, J.; Wu, S.; Bi, W.; Zhao, X.; Liu, G. Z-scheme Fe_2O_3-doped Cu_2O as an efficient photo-Fenton-like catalyst for degradation of phenol. *Mater. Lett.* **2019**, *234*, 13–16. [CrossRef]
43. Zhang, C.; Qin, D.; Zhou, Y.; Qin, F.; Wang, H.; Wang, W.; Yang, Y.; Zeng, G. Dual optimization approach to Mo single atom dispersed g-C_3N_4 photocatalyst: Morphology and defect evolution. *Appl. Catal. B Environ.* **2022**, *303*, 120904. [CrossRef]
44. Bai, J.; Xu, H.; Chen, G.; Lv, W.; Ni, Z.; Wang, Z.; Yang, J.; Qin, H.; Zheng, Z.; Li, X. Facile fabrication of α-Fe_2O_3/porous g-C_3N_4 heterojunction hybrids with enhanced visible-light photocatalytic activity. *Mater. Chem. Phys.* **2019**, *234*, 75–80. [CrossRef]
45. Mo, Z.; Zhu, X.; Jiang, Z.; Song, Y.; Liu, D.; Li, H.; Yang, X.; She, Y.; Lei, Y.; Yuan, S.; et al. Porous nitrogen-rich g-C_3N_4 nanotubes for efficient photocatalytic CO_2 reduction. *Appl. Catal. B Environ.* **2019**, *256*, 117854. [CrossRef]
46. Liu, D.; Li, C.; Ni, T.; Gao, R.; Ge, J.; Zhang, F.; Wu, W.; Li, J.; Zhao, Q. 3D interconnected porous g-C_3N_4 hybridized with Fe_2O_3 quantum dots for enhanced photo-Fenton performance. *Appl. Surf. Sci.* **2021**, *555*, 149657. [CrossRef]
47. Wang, L.; Wang, Y.; Li, X.; He, T.; Wang, R.; Zhao, Y.; Song, H.; Wang, H. 3D/2D Fe_2O_3/g-C_3N_4 Z-scheme heterojunction catalysts for fast, effective and stable photo Fenton degradation of azo dyes. *J. Environ. Chem. Eng.* **2021**, *9*, 105907. [CrossRef]
48. Cui, Y.; Briscoe, J.; Wang, Y.; Tarakina, N.V.; Dunn, S. Enhanced Photocatalytic Activity of Heterostructured Ferroelectric $BaTiO_3$/alpha-Fe_2O_3 and the Significance of Interface Morphology Control. *ACS Appl. Mater. Interfaces* **2017**, *9*, 24518–24526. [CrossRef]

49. Cao, J.; Yang, Z.-H.; Xiong, W.-P.; Zhou, Y.-Y.; Peng, Y.-R.; Li, X.; Zhou, C.-Y.; Xu, R.; Zhang, Y.-R. One-step synthesis of Co-doped UiO-66 nanoparticle with enhanced removal efficiency of tetracycline: Simultaneous adsorption and photocatalysis. *Chem. Eng. J.* **2018**, *353*, 126–137. [CrossRef]
50. Zhou, K.; Liu, Y.; Hao, J. One-pot hydrothermal synthesis of dual Z-scheme BiOBr/g-C_3N_4/Bi_2WO_6 and photocatalytic degradation of tetracycline under visible light. *Mater. Lett.* **2020**, *281*, 128463. [CrossRef]
51. Wang, W.; Zhang, H.; Chen, Y.; Shi, H. Efficient Degradation of Tetracycline via Coupling of Phctocatalysis and Photo-Fenton Processes over a 2D/2D α-Fe_2O_3/g-C_3N_4 S-scheme Heterojunction Catalyst. *Acta Phys. Chim. Sin.* **2022**, *38*, 220100.

coatings

Article

Effects of Nitrogen Doping on Pulling Rate Range of Defect-Free Crystal in CZ Silicon

Chenguang Sun [1,2,3], Zhongshi Lou [3], Xingtian Ai [2,4], Zixuan Xue [3], Hui Zhang [2,4,*] and Guifeng Chen [1,2,*]

1 State Key Laboratory of Reliability and Intelligence of Electrical Equipment, Hebei University of Technology, Tianjin 300132, China; sunchenguang@zhlxsemicon.com
2 School of Materials Science and Engineering, Hebei University of Technology, Tianjin 300132, China; 13363361297@163.com
3 Zhonghuan Advanced Semiconductor Materials Co., Ltd., Yixing 214200, China; louzhongshi@zhlxsemicon.com (Z.L.); xuezixuan@zhlxsemicon.com (Z.X.)
4 Hebei Engineering Laboratory of Photoelectronic Functional Crystals, Hebei University of Technology, Tianjin 300130, China
* Correspondence: zhanghui_1207@163.com (H.Z.); cgfchen@hebut.edu.cn (G.C.)

Abstract: We investigated the effect of nitrogen doping on the pulling rate range of defect-free crystal in silicon with a diameter of 200 mm. It was found that the pulling rate range of defect-free crystal in nitrogen-doped Czochralski silicon is wider and the pulling rate (defect free) is lower than it is in non-nitrogen-doped Czochralski silicon. Under the experiment, the pull rate was from 0.67 mm/min~0.58 mm/min to 0.65 mm/min~0.54 mm/min. To further confirm the above experimental analysis, a numerical simulation process of nitrogen-doped Czochralski and non-nitrogen-doped Czochralski in an industrial system was performed. The V/G value along the S/L interface was the same for both models, but the distribution of Cvi (concentration of vacancy–concentration of self-interstitial) for nitrogen-doped Czochralski crystal silicon was more uniform and flat in a nitrogen-doped single crystal. Furthermore, the nitrogen-doped Czochralski crystal silicon had a smaller void size and a higher oxygen precipitation density. The experimental results are in good agreement with the numerical simulation results.

Keywords: silicon; nitrogen doped; defect free; Cv–Ci; numerical simulation

1. Introduction

In recent years, with the continuous revolution of device design, critical size is becoming smaller, which means that defects from the silicon crystal and wafer process are becoming increasingly stringent. Crystal originated particles (COPs) have become one of the key factors affecting device reliability, while process-induced defects that destroy the integrity of silicon wafers are another key factor. Shang Gao has studied the grinding and lapping-induced surface integrity of silicon wafers and its effects on chemical–mechanical polishing. Through their study, the surface morphology and subsurface damage generated by grinding and lapping were investigated, which helped chemical–mechanical polishing to achieve a polished wafer without foreign particles [1]. At the same time, when the size of COPs is equivalent to the design specification, device yield deterioration is encountered due to gate oxide failure and so on. Therefore, the coordination of COPs size must be considered in the production process, or, in the case of a strict device design, defect-free crystal will be required.

Several types of defects have been found in past years. Yuhong Zhao learned the relationship between the temperature gradient and secondary dendrites, which shows that the temperature gradient is a key factor in maintaining a stable solidification front and interface [2]. Under a stable solidification front and the interface condition, different types of intrinsic point defects (vacancy or self-interstitial) occur during crystal growth based on the V/G theoretical model (where V is the growth rate and G is the near-interface axial temperature gradient) [3]. The vacancy defect is dominant if V/G > the critical value; the vacancies accumulate to become V-defects during the crystal growth process [4]. The self-interstitial defect is dominant when V/G < the critical value; the self-interstitials accumulate to become I-defects after a more severe thermal process. When the V/G value intersects the critical value, the oxidation-induced stacking fault ring defect is prone to occur at this position. The vacancies and self-interstitials reach dynamic equilibrium when V/G is close to the critical value, which can be considered as the pulling rate range of defect-free crystal.

However, nitrogen and other impurities have important effects on the V/G critical value and COP accumulation. They can interplay with oxygen and vacancy to form different kinds of complexes and, furthermore, have a strong influence on the dynamics of microdefects during crystal growth [5]. Doping with donor concentrations above 10^{18} cm^{-3} can lead to both an increase or decrease of 0crit depending on the dopant atom size as the Fermi level effect is smaller than in p-type Si. The trapping of vacancies and the large binding energy is also the reason for the strong suppressing effect of nitrogen doping on the formation of COPs [6]. The impact of N on the vacancy and interstitial silicon intrinsic point defects is very important. Meanwhile, the N implication on the vacancy concentration is larger than it is on the interstitial silicon concentration, and this phenomenon becomes apparent when the N concentration is higher than the specified value [7]. NV (Nitrogen combined with Vacancy) and NO (Nitrogen combined with Oxygen) complexes exist simultaneously in nitrogen-doped Czochralski (NCZ) silicon. Nitrogen–vacancy (NV) and nitrogen–oxygen (NO) complexes compete in the nitrogen-doped Cz-Si crystals and their concentration fraction changes with [OI] value [8]. At present, nitrogen-doped Czochralski single-crystal technology has become popular in semiconductor silicon single-crystal. Infrared absorption experiments on the N-N-O defect in Si have been carried out. The involvement of both N and O in the defect is shown directly by observing N- and O-related isotopic shifts [9]. In the nitrogen-doped and fast-pulled crystals, the nitrogen plays a role in disturbing the growth of vacancy clusters by reducing the vacancy-rich region and the vacancy concentration. Therefore, the vacancy clustering is severely suppressed, and the increased residual vacancy causes anomalous oxygen precipitation [10]. Nitrogen doping could enhance the formation of grown-in oxygen precipitates in larger sizes at high temperatures via the complexes of N_2–V_2–O_X, thus, leading to a significant loss of vacancies before the generation of voids during nitrogen-doped Czochralski crystal growth; on the other hand, the nitrogen doping could also enhance the formation of grown-in oxygen precipitates in smaller sizes at low temperatures via the complexes of Nm–On (Multiple Nitrogen and Oxygen combination). Thus, enhanced by nitrogen doping, the surviving vacancies in relatively lower concentrations resulted in denser voids with smaller sizes in the NCZ crystal [11]. It was believed that nitrogen doping could be an effective pathway for heavily P-doped CZ-Si to enhance oxygen precipitation, thus, increasing the internal gettering capability. Moreover, the nitrogen doping showed an enhancement effect on oxygen precipitation during the prolonged annealing with a rapid thermal processing pretreatment [12]. In nitrogen-doped crystals, the total amount of vacancies composing the voids increased during the crystal growth halt. These residual vacancies can enhance the formation of oxygen-precipitate nuclei during crystal growth [13]. Nitrogen in NCZ silicon enhanced the oxygen precipitation at lower temperatures, while it had no influence on the oxygen precipitation at higher temperatures. Meanwhile, nitrogen interacted with oxygen to form nitrogen–oxygen complexes as heterogeneous nuclei, which enhanced the nucleation of oxygen precipitates [14]. Nitrogen doping in a silicon crystal is known to have two main effects, the reduction in vacancy-related defects and the enhancement of oxygen precipitation. Nitrogen must have an important impact on the process window of COP free crystals based on a large number of research results from international scholars.

In this study, the effect of nitrogen doping on the pulling rate range of defect-free crystal in Czochralski silicon was investigated. A comparison of the formation behavior of various grown-in defects such as crystal-originated particles, and flow pattern defects, which occur in 200 mm diameter Cz-Si crystals grown under gradient pull rate conditions, with and without nitrogen doping, respectively, was performed.

At the same time, the V/G value of the S/L interface, Cvi (difference between voids and self-interstitials concentration), voids size, etc., were analyzed by numerical simulation and compared with the experimental results. Based on this study, the effect of nitrogen doping on the pulling rate range of defect-free crystal in Czochralski silicon can be discussed effectively.

2. Experimental Section

2.1. Experimental Details

In this experiment, a conventional 8-inch Czochralski single-crystal growth furnace was used, and the Czochralski furnace type was TDR110. To investigate the effects of nitrogen doping on the pulling rate range of defect-free crystal, two 200 mm (100) p-type silicon crystals were grown under almost the same conditions, besides nitrogen concentration, as shown in Table 1. The body length of the experimental single crystal was 700 mm, which was grown in a graphite thermal field of 24 inches, and the feeding volume was 180 kg.

Table 1. Key experimental conditions of two crystals.

Crystal Type	Nitrogen Concentration (Atoms/cm^3)	Pulling Rate (mm/min)	Diameter, mm
CZ	0	0.67→0.5	200
NCZ	1.9×10^{13}~3×10^{13}	0.67→0.5	200

As the pulling speed gradually changed from high to low, the crystal defect type mainly went through three stages, from V-rich (vacancy dominance) to V–I equilibrium (defect free), and then to I-rich (interstitial silicon dominance). Therefore, the experiment of gradually varying the pulling speed was performed to find the changes in the defect-free crystal (V–I equilibrium region) of non-nitrogen-doped Czochralski (CZ) silicon and nitrogen-doped Czochralski (NCZ) silicon, respectively. The pulling rate of the two crystals changed from 0.67 mm/min to 0.5 mm/min, as shown in Figure 1.

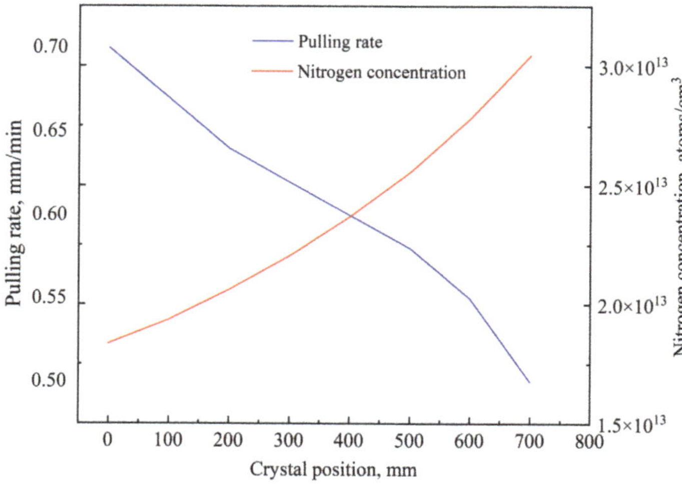

Figure 1. Experimental crystal pulling rate and nitrogen concentration.

To obtain NCZ silicon, the silicon polish wafer with a silicon nitride epitaxial layer and polysilicon raw material was loaded into the quartz crucible. Then, nitrogen entered into the single-crystal silicon via the impurity segregation effect during the crystal growth process. The distribution of nitrogen was different in different crystal positions, which varied from low to high due to its low segregation coefficient, as shown in Figure 1. The nitrogen concentration in the grown crystal was calculated using the segregation coefficient of 0.0007 as it was difficult to measure the low nitrogen concentration in the silicon wafer. The common nitrogen concentration levels of the crystal growth in this experiment were approximately 1.9×10^{13} atoms/cm^3 and 3×10^{13} atoms/cm^3 at the starting position and the ending position, respectively.

2.2. Experimental Results and Discussions

2.2.1. Copper Decoration Analysis

The properties of monocrystalline silicon are divided into three main categories depending on the pulling speed, including V-rich (vacancy dominance), V–I equilibrium (defect free), and I-rich (interstitial silicon dominance). Chemical etching is widely applied to decorate microdefects in monocrystalline silicon to qualitatively distinguish the crystal defect types. A general copper decoration technique has been widely used for qualitative analyses of crystal microdefect distribution, which is identified by saturating the wafer with copper at a high temperature followed by rapid cooling, allowing copper precipitate growth, followed by surface polishing and subsequent microdefect-decorating etching. In this experiment, the difference between CZ and NCZ defect-free windows was verified using copper decoration technology.

First, longitudinal sections were obtained from the CZ and NCZ silicon crystals by cutting along the axis. To obtain samples with a flat surface, the process of grinding and polishing was successively passed. Then, both groups of sample defects were tested using the cooper decoration technique [15,16]. The following steps were followed sequentially—(a) the two samples were coated uniformly on the back with high-concentration cupric nitrate ($Cu(NO_3)_4 \cdot 5H_2O$); (b) the two samples with copper nitrate were annealed at 900 °C for 10 min; (c) rapid cooling of the silicon wafer was performed to allow the heterogeneous nucleation of copper on the microdefects and subsequent growth of copper precipitate colonies; (d) chemical surface polishing was performed to clean the wafer surface; and (e) decorating etching was performed to decorate the copper precipitate colonies.

In the CZ silicon crystal, R-1 represents the near-defect-free crystal region, R-2 represents the defect-free crystal region, R-3 represents the low-density self-interstitial crystal region, and R-4 represents the I-rich crystal region. In NCZ silicon crystals, due to the influence of nitrogen on the distribution of defects, R-1 changed from near-defect-free to a V-rich crystal region, no change was observed in R-2, R-3 transformed from low-density self-interstitial to a defect-free crystal region, and R-4 represented a smaller I-rich region. As a result, the NCZ silicon crystals had a wider region of defect-free crystals.

According to Voronkov's V/G (growth rate/temperature gradient), the type of point defect in a silicon crystal transforms from a vacancy to a self-interstitial atom as the growth rate decreases. For CZ silicon crystal, no V-defects are in the R-1 or R-2 areas with a pulling rate from 0.67 mm/min to 0.58 mm/min. I-defects are found when the pulling rate is below 0.58 mm/min, and dislocation loops or dislocation loop clusters form in R-3 through self-interstitial aggregation, which are identified as A-defects. Therefore, the pulling rate range of perfect crystal for CZ is from 0.67 mm/min to 0.58 mm/min, as shown in Figure 2.

For NCZ silicon crystal, the near-defect-free crystal is transformed into V-rich crystal in the R-1 region, and the I-rich crystal is transformed into a defect-free single crystal in the R-3 region. There are no V-defects in the R-2 and R-3 regions where the pulling rate is from 0.65 mm/min to 0.54 mm/min. I-defects are found when the pulling rate is below 0.54 mm/min, and dislocation loops or dislocation loop clusters form in R-3 through self-interstitials aggregation, which are identified as A-defects. Therefore, the pulling rate range of perfect crystal for NCZ is from 0.65 mm/min to 0.54 mm/min.

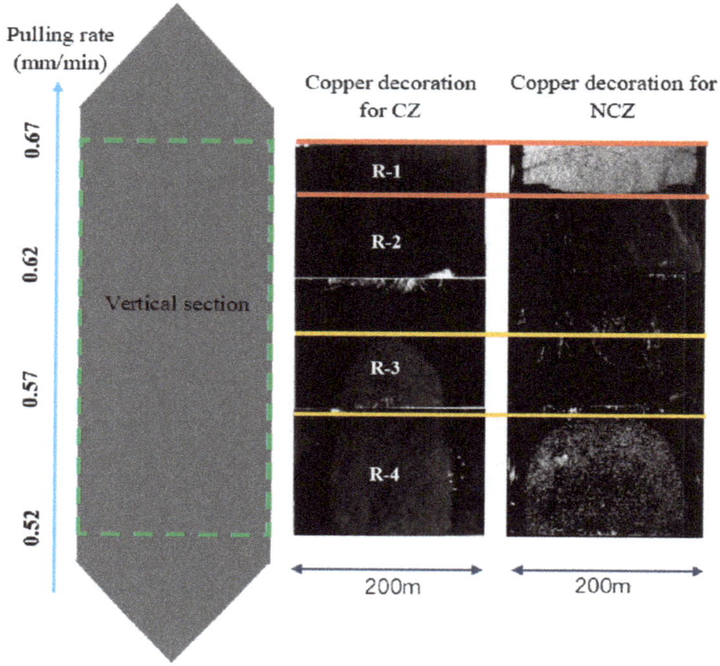

Figure 2. Cooper decoration mapping of longitudinal sections for CZ and NCZ crystal.

2.2.2. COPs and FPDs Analysis of Polish Wafer

To compare and analyze the results of copper decoration, experimental single-crystal ingots of CZ and NCZ were grown according to the original experimental boundary condition. The same position of the new experimental single-crystal ingots was then processed into a polish wafer. On the one hand, the particle distributions of the experimental CZ and NCZ polish wafers were tested using an SP2 device. On the other hand, the particle type (COPs or large dislocation particles (LDPs)) was tested using the flow pattern defect (FPD) technique [17,18] such that the samples were vertically immersed and etched in the Secco etchant (50%HF:0.15 mol/L$K_2Cr_2O_7$ = 2:1) after polishing treatment. The morphology of the FPDs was observed using an optical microscope. The COP distributions of R-1, R-2, R-3, and R-4 for the CZ and NCZ silicon polish wafers were tested using the SP2 device. The distribution of COPs is shown in Table 2. The FPDs results map is shown in Figure 3.

Table 2. COPs inspected by SP2 at 0.065 μm mapping for CZ and NCZ experiments.

	R-1	R-2	R-3	R-4
COP Distribution at 0.065 μm (CZ)	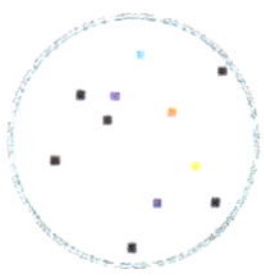	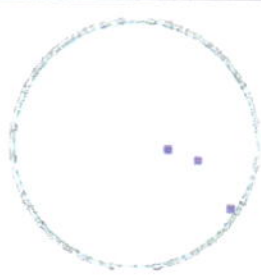	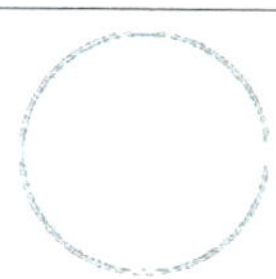	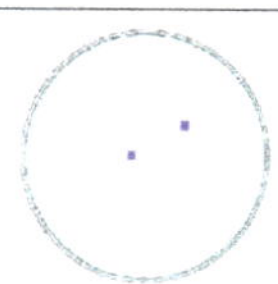
COP Distribution at 0.065 μm (NCZ)	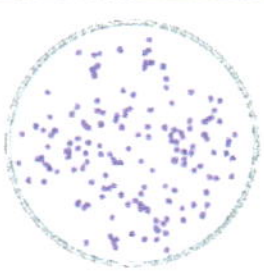	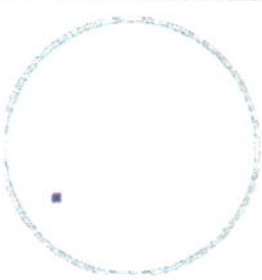	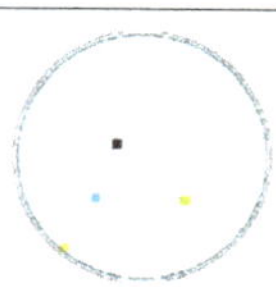	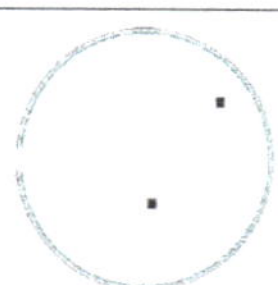

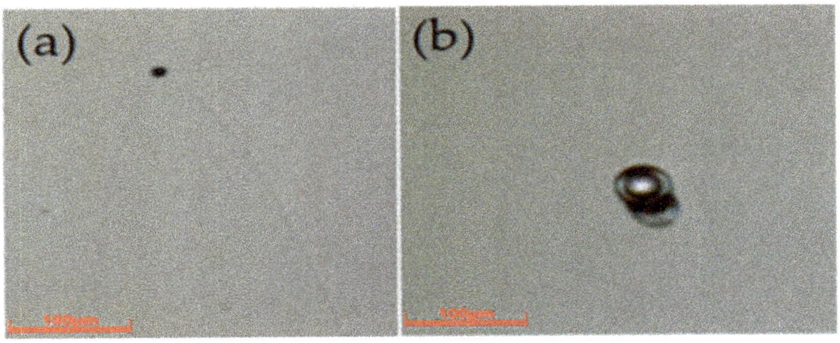

Figure 3. COPs (**a**) and LDPs (**b**) using FPD method.

Table 2 shows the COPs at 0.065 μm distribution maps detected by SP2 device. There are scattered COPs in the R-1 region from the CZ silicon crystal polish wafer. It was found that there are some agminated FPDs in the center of the R-1 samples from the CZ silicon polish wafer, like those shown in Figure 3a. The tip of the FPD is void, which could certify that these scattered particles are corresponding COPs. There are no FPDs on the edge of the CZ silicon polish wafer samples. The R-1 area should be a near-defect-free zone combined with the analysis of copper decoration results. At the same time, there are no COPs for the R-2, R-3, and R-4 regions, other particles should be foreign particles from the polishing process. There is no FPD, but there are some etch pits in the center of the R-3 and R-4 samples from the CZ silicon polish wafers. These etch pits are oval or twin etched pits from the FPD results of the R-3 and R-4 regions, like those shown in Figure 3b. These etch pits from the FPD test of R-3 and R-4 samples are LDPs (I-defects) which are in good agreement with the copper decoration results above and the numerical simulation results below.

There are a large number of COPs clustered in the central area in the R-1 region from the NCZ silicon crystal wafer. It was found that there are a large number of centrally aggregated FPDs in the center of the R-1 samples from the NCZ silicon polish wafers, like those shown in Figure 3a. Similar to the FPDs of the R-1 samples from the CZ silicon polish wafer samples, the tip of the FPD is void, which could certify that these scattered particles are also corresponding COPs. Meanwhile, there are no COPs for the R-2, R-3, and R-4 regions. Strangely, FPD test results showed that there are no FPDs or etch pits in the R-3 region of the NCZ samples, which is different from the FPD test results of the NCZ samples. Combined with the copper decoration results analysis, we believe the R-3 zone changed from an I-defect zone to a defect-free zone after nitrogen doping . However, etch pits (LDPs) were still found in the R-4 region of the NCZ samples, like those shown in Figure 3b. The distribution of COPs and LDPs is in good agreement with the results of the copper decoration analysis.

3. Numerical Simulation Section

3.1. Numerical Simulation Setup

The agreement of CGSim simulations with experimental results has been extensively verified [19–21]. In this work, the CGSim (crystal growth simulator) program from STR Inc., which is based on the finite volume method (FVM), was used for all computations. Two-dimensional and three-dimensional steady-state simulations were performed with the basic module and flow module of the CGSim software package for computations of the global heat and mass transfer in the entire furnace, taking into account the turbulent convection in the melt and gas regions, the chemical mode of nitrogen doping, and the dynamics of intrinsic point defects in the crystal. The process of numerical simulation is mainly divided into three parts—(1) build the properties and formulas of the model filling materials and define each boundary combined with the process parameters; (2) construct the required crystal position (CP) locations, divide the grids, and adjust the encryption of the structured and unstructured grids and laminar flow grids; and (3) set the GUI operation, set the model solving conditions, add the required chemicals, and define the process parameter variables. The grid creation mainly consists of two modules—automatic grid division and manual grid modification. First, we used the automatic grid division function to automatically divide gas, solid, and liquid grids and generate unstructured grids. Then, we used the manual grid modification function to modify the solid–liquid interface of crystal and melt square area, we also modified the heat source conduction part into structured grids.The result is shown in Figure 4. Finally, the small mass and heat transfer space, crystal melt phase transition interface, gas–liquid intersection interface, and gas–solid boundary interface were mesh-corresponded. The problem of crystallization front correction requires a special grid design—the melt and the crystal should have a block with a quadrangular grid, which contains the melt–crystal interface. Boundary conditions should be set on different tabs separately; radiation, gas, and solid tabs should be used. External and internal subtabs are for external and internal boundaries, respectively. A special subtab with gas is proposed for boundary conditions to be set on boundaries between solid blocks and gas. Melt density (RHO) is supposed to be temperature-dependent to take into account the natural convection of the melt. At the same time, the defect simulation analysis is based on the steady-state heat and mass transfer simulation, and the whole crystal region needs to be structured with mesh encryption to ensure that the defect iteration can converge. Here, we discuss the effects of nitrogen doping on the distributions of the V and I concentrations and the V/G value along the S/L interface in a growing Si crystal.

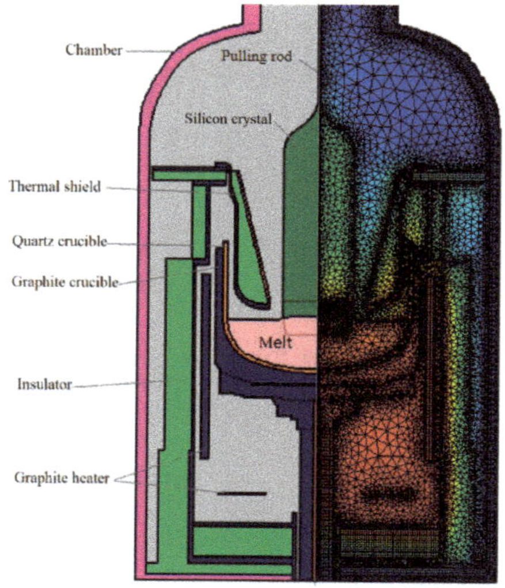

Figure 4. Material arrangement (**left**) and computational grids (**right**) in a furnace.

3.2. *V/G Value Based on the Different Growth Rates*

To further confirm the above experimental analysis, we carried out a numerical simulation analysis using three pulling rates, 0.67 mm/min, 0.65 mm/min, and 0.55 mm/min, at the crystal position of 600 mm using simulation software CGSim. As shown in Figure 5, the V/G value in the center and edge of the growth interface was smaller under the three conditions of pulling rate. The forced convection below the solid–liquid interface makes the temperature gradient of the solution at the center larger, resulting in a small V/G value at the center. Because the edge is close to the gas–solid–liquid triple point, the temperature loss there is large. Therefore, the temperature gradient at the edge surges, resulting in a smaller V/G value at the edge. When the pulling speed is 0.55 mm/min, the V/G value in the center is less than the critical value of 0.013 $cm^2/min*K$ [22], which proves that the self-interstitial silicon may appear in the center when the pulling rate is gradually reduced, which agrees with the experiment. When the pulling speed is 0.7 mm/min, all the V/G values except for the edge are higher than the critical value of 0.013 $cm^2/min*K$. Meanwhile, COP defects appear on the wafer when the pulling speed continues to increase, which is in good agreement with the experimental results. When the pulling speed is 0.65 mm/min, except for the edge, the remaining V/G values are close to the critical value of 0.013 $cm^2/min*K$. When the V/G value is close to the critical value of 0.013 $cm^2/min*K$, we believe that the concentration of self-clearance and vacancy defects generated by the solid–liquid interface is the same. At this time, most of the self-clearance and vacancy can cancel each other through diffusion and recombination during crystal growth, so that the growth of defectless single crystals can be realized.

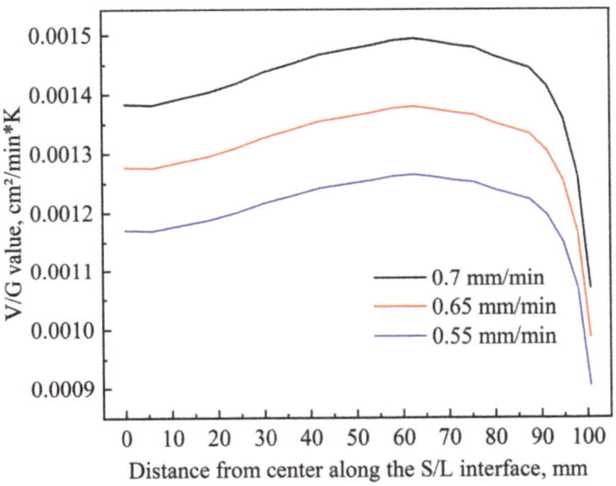

Figure 5. Distribution of V/G for different pulling rates.

3.3. Defects for CZ and NCZ Crystal

There is a significant difference between the center and edge from the distribution of Cvi (concentration of vacancy–concentration of self-interstitial) for CZ crystal, as shown in Figure 6. Muhammad established six types of vacancy models in CrSI semiconductor monolayer [23], which helps us comprehend all the possibilities that vacancies can form and recombine into during the CZ crystal growth process. The internal temperature distribution of NCZ and CZ crystals is the same under the same pulling conditions. NCZ crystal promotes COP nucleation, resulting in an increase in COP density, but a decrease in COP size. As can be seen from Figure 6, when T = 1450 K, the center Cvi of NCZ crystal is 1.0715×10^{13} ($1/cm^3$), which is significantly larger than the center Cvi of 1.0072×10^{13} ($1/cm^3$) of CZ crystal. At the same time, more vacancies remain in the NCZ crystal, and these excess residual vacancies recombine with self-interstitial. As can be seen from Figure 6, when T = 1360 K, the edge Cvi of NCZ crystal is 9.5489×10^{12} ($1/cm^3$), which is significantly smaller than the center Cvi of 1.0693×10^{13} ($1/cm^3$) of CZ crystal. The simulation results are in good agreement with the experimental results.

At the same time, the influence of different dopants on crystal particles is also different. Some dopants inhibit the growth of particles and increase their density, while some dopants increase the size of particles and reduce their density. The size of voids for NCZ crystal silicon is smaller than that for CZ crystal silicon, and its pore size is reduced by 15–20 nm, as shown in Figure 7. In NCZ crystal silicon, the combination of nitrogen, oxygen, and vacancy to form a bulk defect, will reduce the concentration of free vacancy, not only inhibiting the growth of COPs but also reducing the size of COPs. At the same time, according to the Cvi distribution in Figure 6, the Cvi near the edge is more convex, which means that the vacancy concentration in this part is larger and may generate larger voids, which is in good agreement with the void size distribution shown in Figure 7.

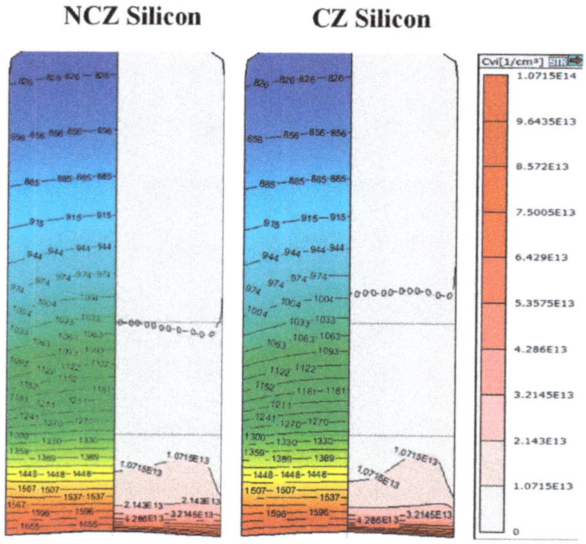

Figure 6. Cvi (Cv–Ci) distribution of CZ and NCZ crystal.

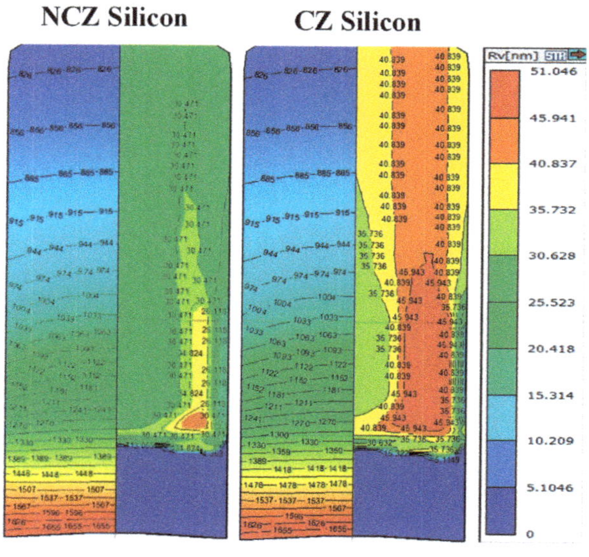

Figure 7. Rv (V-defects size) of CZ and NCZ crystal.

The density of oxygen precipitation in NCZ crystal silicon increased significantly and was higher than it was in CZ silicon, and the overall oxygen precipitate density increased by 0.5×10^9 to 2×10^9, as shown in Figure 8. The density of oxygen precipitates in the center of the NCZ crystal is approximately 2.2161×10^9–4.4322×10^9 ($1/cm^3$), which is slightly higher than the concentration of oxygen precipitates in the center of the CZ crystal, 2.0816×10^9–3.4694×10^9 ($1/cm^3$). At the same time, the density of oxygen precipitates at the edge of the NCZ crystal is approximately 2.9548×10^9–6.6483×10^9 ($1/cm^3$), which is significantly higher than the concentration of oxygen precipitates at the center of the CZ crystal, 2.7755×10^9–4.3945×10^9 ($1/cm^3$).

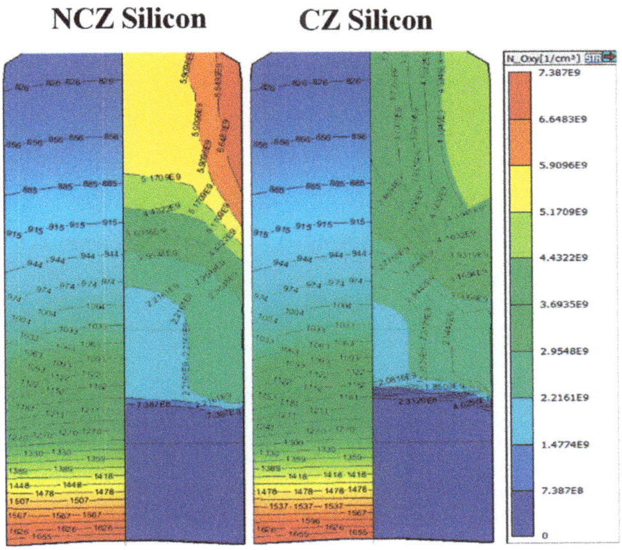

Figure 8. Density of oxygen precipitation for CZ and NCZ crystal.

Nitrogen doping promotes the nucleation of oxygen precipitation due to the decrease in vacancy supersaturation but does not promote the growth of the void. Therefore, the size of the void decreases and the density increases. Excess vacancies, which remain in the crystals just after void formation, will recombine with interstitial silicon atoms or promote oxygen precipitation so that interstitial silicon atoms are reduced [24].

Through the comparative analysis of numerical simulation results, we believe that NCZ crystal silicon has better Cvi distribution than CZ crystal silicon, which is conducive to the diffusion and recombination of intrinsic point defects and further crystal improvement. At the same time, due to the effect of nitrogen on vacancy supersaturation, it can inhibit the production of large COPs. Because more vacancy remains in the growth process of NCZ crystal silicon than CZ crystal silicon, the concentration of oxygen precipitates inside the crystal can be increased to improve the internal impurity absorption capacity and mechanical strength of the chip to a certain extent.

4. Conclusions

We investigated the effect of nitrogen doping on the pulling rate range of defect-free crystal in Czochralski (CZ) silicon. It was suggested that the pulling rate range of defect-free crystal silicon is from 0.67 mm/min~0.58 mm/min to 0.65 mm/min~0.54 mm/min. The near-defect-free crystal is transformed into V-rich in the R-1 region, and the I-rich crystal is transformed into a defect-free single crystal in the R-3 region. The simulation results show that the V/G value along the S/L interface is the same for NCZ and CZ models, but the distribution of Cvi (Cv-Ci) for NCZ crystal silicon is more uniform and flat in a nitrogen-doped single crystal. The size of voids in NCZ crystal silicon is smaller than that in CZ crystal silicon. The density of oxygen precipitation in NCZ crystal silicon increased significantly and was higher than it is in CZ silicon. The experimental results are in good agreement with the numerical simulation results. Based on the above results, on the one hand, nitrogen doping could enhance the formation of voids during crystal growth and residual vacancies could remain, on the other hand, the residual vacancies recombine with interstitial silicon or promote oxygen precipitation so that interstitial silicon atoms are reduced. The research in this paper provides some guidance and suggestions for the production of defect-free single-crystal technology with a high demand for internal gettering in the semiconductor silicon material industry.

Author Contributions: Formal analysis, Z.L., Z.X. and G.C.; Data curation, Z.L.; Writing—original draft, C.S., X.A., H.Z. and G.C.; Writing—review & editing, H.Z., Z.L. and G.C. All authors have read and agreed to the published version of the manuscript.

Funding: This work was supported by the project for the Science and Technology Correspondent of Tianjin City (Grant No. 20YDTPJC01710) and the Research Foundation of Education Bureau of Hebei (Grant No. QN2021044).

Institutional Review Board Statement: Not applicable.

Informed Consent Statement: Not applicable.

Data Availability Statement: Not applicable.

Conflicts of Interest: The authors declare no conflict of interest.

References

1. Gao, S.; Li, H.; Huang, H.; Kang, R. Grinding and Lapping Induced Surface Integrity of Silicon Wafers and Its Effect on Chemica Mechanical Polishing. *Appl. Surf. Sci.* **2022**, *599*, 153982. [CrossRef]
2. Zhao, Y.; Zhang, B.; Hou, H.; Chen, W.; Wang, M. Phase-Field Simulation for the Evolution of Solid/Liquid Interface Front ir Directional Solidification Process. *J. Mater. Sci. Technol.* **2019**, *35*, 1044–1052. [CrossRef]
3. Voronkov, V. The mechanism of swirl defects formation in silicon. *J. Cryst. Growth* **1982**, *59*, 625–643. [CrossRef]
4. Voronkov, V. Vacancy and self-interstitial concentration incorporated into growing silicon crystals. *J. Appl. Phys* **1999**, *86* 5975–5982. [CrossRef]
5. Yu, X.; Chen, J.; Ma, X.; Yang, D. Impurity engineering of Czochralski silicon. *Mater. Sci. Eng. R Rep.* **2013**, *74*, 1–33. [CrossRef]
6. Vanhellemont, J.; Kamiyama, E.; Sueoka, K. Silicon Single Crystal Growth from a Melt: On the Impact of Dopants on the v/C Criterion. *ECS J. Solid State Sci. Technol.* **2013**, *2*, P166–P179. [CrossRef]
7. Taniguchi, M.; Sueoka, K.; Hourai, M. Density Functional Theory Study on Concentration of Intrinsic Point Defects in Growing N-Doped Czochralski Si Crystal. *J. Cryst. Growth* **2021**, *571*, 126249. [CrossRef]
8. Kajiwara, K.; Torigoe, K.; Harada, K.; Hourai, M.; Nishizawa, S. Oxygen concentration dependence of as-grown defect formatior in nitrogen-doped Czochralski silicon single crystals. *J. Cryst. Growth* **2021**, *570*, 126236. [CrossRef]
9. Berg Rasmussen, F.; Öberg, S.; Jones, R.; Ewels, C.; Goss, J.; Miro, J.; Deák, P. The nitrogen-pair oxygen defect in silicon. *Mater. Sci Eng.* **1996**, *B36*, 91–95. [CrossRef]
10. Park, B.M.; Seo, G.H.; Kim, G. Nitrogen-Doping Effect in a Fast-Pulled Cz-Si Single Crystal. *J. Cryst. Growth* **2001**, *222*, 74–81 [CrossRef]
11. Yu, X.; Yang, D.; Ma, X.; Yang, J.; Li, L.; Que, D. Grown-in Defects in Nitrogen-Doped Czochralski Silicon. *J. Appl. Phys.* **2002**, *92* 188–194. [CrossRef]
12. Zeng, Y.; Chen, J.; Ma, M.; Wang, W.; Yang, D. Enhancement effect of nitrogen co-doping on oxygen precipitation in heavily phosphorus-doped Czochralski silicon during high-temperature annealing. *J. Cryst. Growth* **2009**, *311*, 3273–3277. [CrossRef]
13. Umeno, S.; Ono, T.; Tanaka, T.; Asayama, E.; Nishikawa, H.; Hourai, M.; Katahama, H.; Sano, M. Nitrogen Effect on Grown-in Defects in Czochralski Silicon Crystals. *J. Cryst. Growth* **2002**, *236*, 46–50. [CrossRef]
14. Yang, D.; Ma, X.; Fan, R.; Zhang, J.; Li, L.; Que, D. Oxygen precipitation in nitrogen-doped Czochralski silicon. *Phys. B Condens Matter* **1999**, *273–274*, 308–311. [CrossRef]
15. Mule'Stagno, L. A Technique For Delineating Defects in Silicon. *Solid State Phenom.* **2002**, *82–84*, 753–758. [CrossRef]
16. Kulkarni, M.S.; Libbert, J.; Keltner, S.; Muléstagno, L. A Theoretical and Experimental Analysis of Macrodecora-tion of Defects in Monocrystalline Silicon. *J. Electrochem. Soc.* **2002**, *149*, G153. [CrossRef]
17. Zhang, J.; Liu, C.; Zhou, Q.; Wang, J.; Hao, Q.; Zhang, H.; Li, Y. Evolution of Flow Pattern Defects in Bo-ron-Doped <100> Czochralski Silicon Crystals during Secco Etching Procedure. *J. Cryst. Growth* **2004**, *269*, 310–316. [CrossRef]
18. Xu, W.; Chen, J.; Ma XYang, D.; Gong, L.; Tian, D. Flow pattern defects in germanium-doped Czochralski silicon crystals. *Appl. Phys. A* **2011**, *104*, 349–355. [CrossRef]
19. Chen, J.-C.; Chiang, P.-Y.; Nguyen, T.H.T.; Hu, C.; Chen, C.-H.; Liu, C.-C. Numerical Simulation of the Oxygen Concentration Distribution in Silicon Melt for Different Crystal Lengths during Czochralski Growth with a Trans-verse Magnetic Field. *J. Cryst. Growth* **2016**, *452*, 6–11. [CrossRef]
20. Mukaiyama, Y.; Sueoka, K.; Maeda, S.; Iizuka, M.; Mamedov, V.M. Numerical Analysis of Effect of Thermal Stress Depending on Pulling Rate on Behavior of Intrinsic Point Defects in Large-Diameter Si Crystal Grown by Czochralski Method. *J. Cryst. Growth* **2020**, *531*, 125334. [CrossRef]
21. Nguyen, T.H.T.; Chen, J.C.; Hu, C.; Chen, C.H. Numerical Simulation of Heat and Mass Transfer during Czo-chralski Silicon Crystal Growth under the Application of Crystal-Crucible Counter- and Iso-Rotations. *J. Cryst. Growth* **2019**, *507*, 50–57. [CrossRef]
22. Von Ammon, W.; Dornberger, E.; Oelkrug, H.; Weidner, H. The Dependence of Bulk Defects on the Axial Tem-perature Gradient of Silicon Crystals during Czochralski Growth. *J. Cryst. Growth* **1995**, *151*, 273–277. [CrossRef]

23. Muhammad, I.; Ali, A.; Zhou, L.; Zhang, W.; Wong, P.K.J. Vacancy-Engineered Half-Metallicity and Magnetic Anisotropy in CrSI Semiconductor Monolayer. *J. Alloys Compd.* **2022**, *909*, 164797. [CrossRef]
24. Von Ammon, W.; Hölzl, R.; Virbulis, J.; Dornberger, E.; Schmolke, R.; Gräf, D. The Impact of Nitrogen on the Defect Aggregation in Silicon. *J. Cryst. Growth* **2001**, *226*, 19–30. [CrossRef]

coatings

MDPI

Article

Investigation of Oxygen Behavior under Different Melt Flow, Diffusion Boundary Layer, and Crystal-Melt Interface in a 300 mm Silicon Crystal Growth with Cusp Magnetic Field

Chenguang Sun [1,2,3], Xingtian Ai [2,4], Hui Zhang [2,4,*], Hungpang Chou [3], Huiyun Lyu [3] and Guifeng Chen [1,2,*]

1 State Key Laboratory of Reliability and Intelligence of Electrical Equipment, Hebei University of Technology, Tianjin 300132, China; sunchenguang@zhlxsemicon.com
2 School of Materials Science and Engineering, Hebei University of Technology, Tianjin 300132, China; 13363361297@163.com
3 Zhonghuan Advanced Semiconductor Materials Co., Ltd., Yixing 214200, China; hpchou@zhlxsemicon.com (H.C.); lvhuiyun@zhlxsemicon.com (H.L.)
4 Hebei Engineering Laboratory of Photoelectronic Functional Crystals, Hebei University of Technology, Tianjin 300130, China
* Correspondence: zhanghui_1207@163.com (H.Z.); cgfchen@hebut.edu.cn (G.C.)

Abstract: The silicon single crystals for semiconductor application are usually grown by the Czochralski (CZ) method. In this paper, we studied a 300 mm Czochralski silicon crystal grown with a cusp magnetic field to be used for an insulated gate bipolar transistor (IGBT). Different positions of the zero-Gauss plane (ZGP) under a cusp magnetic field were simulated and compared to numerical analysis. We investigated three factors that affected the oxygen concentration in the crystal, including (1) melt convection, (2) melt flow velocity near the quartz crucible wall, and (3) the diffusion boundary layer. We also studied the shape of the solid/liquid interface at the same time. The simulation results show that a change in the ZGP of the cusp magnetic field (CMF) strongly affects the convection in the melt, which leads to a difference in the thickness of the boundary layer near the wall of the quartz crucible. We investigated the relationship of the ZGP, convection in the melt, and the thickness of the boundary layer. In this way, we determined how to reduce oxygen diffusing into the melt and finally into the crystal. After simulation results were obtained, we pulled single crystals under the three configurations. The results show that the experimental data of the oxygen content and shape of the solid/liquid interfaces are consistent with the simulation results.

Keywords: zero-Gauss plane; cusp magnetic field (CMF); crystal-melt interface; magnetic Czochralski method

1. Introduction

In recent decades, with the rapid development of electronic components, massive business opportunities have emerged in the industrial power control, automotive, photovoltaic, and wind-electrical fields. It is well known that the float-zone (FZ) method is used to make silicon wafer substrate for insulated gate bipolar transistors (IGBTs). To reduce costs and raise productivity, a more cost-effective 300 mm diameter wafer substrate can be used for IGBTs. Therefore, silicon material grown by the CZ process with magnetic field equipment can be used as an IGBT substrate. IGBTs have been widely applied and are mainly used in high-voltage, high-current, and high-power scenarios. Compared with chip performance, users pay more attention to the stability and reliability of the products, which require a relatively low level of impurities in the crystal. The oxygen- and carbon-relative compounds formed after heat treatment by oxygen and carbon atoms in the crystal may impact the device's performance [1].

Further studies show that the oxygen concentration of IGBT devices is required to be as low as possible. Since the excessive oxygen will be precipitated into SiO_x during the heat treatment process of the device manufacturing process, the compounds would deteriorate the recombination lifetime of IGBTs [2]. In addition to that, when silicon wafers containing excessive oxygen are subjected to the low temperature of about 450 °C during crystal cooling or process cooling, oxygen thermal donors will form, resulting in changes in the electrical properties of the substrate, which is one of the important quality characteristics, requiring uniformity and stability. In particular, it is required that the resistivity distribution in the crystal is uniform, and the resistivity does not change after heat treatment in the device manufacturing process. The oxygen concentration and its uniformity in the CZ crystal mainly depend on the flow state in the molten liquid and some process parameters of crystal pulling. With the increase in the diameter of a silicon single crystal, the number of silicon crystal growth systems in the molten liquid increases, and the heat convection in the crucible becomes more intense [3]. To ensure the quality of crystallization, the heat convection in the molten liquid must be suppressed. Applying a magnetic field effectively controls the liquid flow during silicon growth [4].

The magnetic fields Czochralski (MCZ) technique is capable of growing low oxygen, low micro defects, and high resistivity CZ silicon. Beginning in 1980, various types of magnetic fields were developed, including magnetic field orientation (vertical magnetic fields CZ (VMCZ) for longitudinal magnetic fields, horizontal magnetic fields CZ (HMCZ) for transverse magnetic fields and cusp-shaped magnetic fields) [5], and crystal growth studies were conducted. Kakomoto [6] studied the VMCZ magnet and found that the VMCZ magnet field could suppress the radial heat convection, and oxygen concentration decreased with an increase in the longitudinal magnetic field intensity. Liu's model showed that near the crucible wall, HMCZ inhibited the longitudinal molten flow caused by heat convection, delaying and blocking the oxygen transfer [7]. However, the radial magnetic field destroyed the symmetry of the growth system relative to the growth axis, so the effect of the magnetic field was also non-axisymmetric, resulting in the appearance of growth fringes at the crystal surface intervals. During the growth of VMCZ, the oxygen concentration in the crystal is affected by many variables and cannot be easily controlled. Therefore, in this research, the cusp magnetic field is considered as a more suitable method for crystal growth to better control the oxygen concentration of silicon single crystals than HMCZ and VMCZ. In past studies, the oxygen concentration of crystals can be improved by the application of the magnetic field and by changing the crucible rotation, but simply changing the crucible rotation parameter is not sufficient to maintain the oxygen level in the ideal range. In Borisov [8], a steady-state turbulence model was proposed for the extended hypothesis of Reynolds number stress tensor, which takes into account the different mechanisms of Reynolds number stress anisotropy and can be used for fast two-dimensional engineering calculations. Nguyen [9] numerically investigated the effects of a balanced/unbalanced cusp magnetic field (CMF) and crystal crucible inversion/iso-rotation on the heat, flow, and oxygen distribution during the growth of 8-inch silicon crystals. In Gunjal [10], a global model was established to study the effect of crystal and crucible rotation on melt flow in the absence and presence of electromagnetic fields.

The silicon melt is conductive, and under the action of a magnetic field, the flow of the melt will inevitably cause an induced current and thus generate Lorentz force. Under the action of Lorentz force, the heat convection in the silicon melt is inhibited, and the oxygen, point defects, and other impurities in the silicon crystal are suppressed. The oxygen vacancies are the most common point defects in technologically important complex oxide materials and influence their physical properties [11]. Proper distribution of the magnetic field can improve the uniformity of a single crystal and reduce initial oxygen concentration, boron, aluminum, and other impurities from the quartz crucible into the melt, thus improving the quality of silicon. The experimental and theoretical studies show that using a cusp magnetic field to pull crystals can control molten liquid's convection, impurities transport, and defects formation [12]. The cusp magnetic field is asymmetrically distributed relative to the pulling shaft, and the magnetic fields cancel each other at the center point, resulting in zero magnetic field strength in the vertical direction. A vertical magnetic field is formed away from the center of the magnetic field, and a horizontal magnetic field is formed along the radius. In this way, by applying a cusp magnetic field to the molten silicon liquid, the convection of the molten silicon liquid along the direction orthogonal to the magnetic field line can be controlled. As a result, it can not only reduce the amount of oxygen from the quartz crucible but also promote oxygen evaporation in the silicon molten liquid to stably grow a single crystal with lower oxygen concentration.

2. Experimental Setup and Modeling Approach

An installation schematic of the industrial CZ system with the cusp magnet is shown in Figure 1. The system mainly consists of a single crystal furnace, heater, graphite crucible, quartz crucible, crystal pulling, and rotation equipment. The system is applied to grow a 300 mm diameter silicon single crystal; the initial weight of the silicon melt is 260 kg, and the single crystal growth parameters are shown in Table 1. Notably, the crystal and crucible rotations are opposite. Argon flow rates and furnace pressure are fixed during crystal growth. The CMF is added to the outside of the single crystal furnace, which is specially produced at Zhonghuan Advanced Semiconductor Materials Co., Ltd., Yixing, China. The coil of the CMF is coaxial with the crystal growth furnace; there are two coils in the cusp-magnetic configuration, upper and lower coils, which are used to induce a CMF, and the current of the upper and lower coils is opposite. The zero-Gauss plane (ZGP) is determined by the Gauss meter and is at the mid-position between the upper and lower coils in the CMF configuration, as shown by the blue line. The reference intensity of the magnetic field is 0.09 T, which is set at the intersection of the ZGP position and the quartz crucible. The intensity remains constant during the growth process.

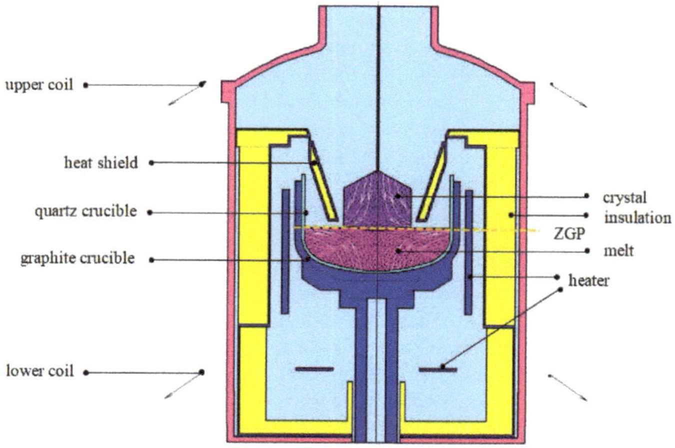

Figure 1. Schematic diagram of the major components of the industrial CZ furnace.

Table 1. Single crystal growth parameter.

Parameter	Value
Furnace pressure, Pa	2000
Argon gas flow rate, slpm (standard liter per minute)	95
Crystal pulling rate, mm/h	24.5
The ratio of crystal to crucible rotation rate	6:−1
Magnetic field strength, T	0.09

3. Description of Cusp Magnetic Field

We conducted numerical simulations on the control of the oxygen concentration using a CGSim (crystal growth simulator). The CMF is shown in Figure 2. During crystal growth, the crystal solid-liquid interface is fixed in the symmetric region of two coils. The effective magnetic field is composed of a longitudinal magnetic field perpendicular to the bottom of the crucible and a radial magnetic field perpendicular to the wall of the crucible, while there is no orthogonal magnetic field composition on the surface of the molten liquid. The position of the ZGP is determined by changing the position of the fixed coil while maintaining a constant magnetic ratio (MR) of the upper and lower coils, which means the upper and lower coils have the same distribution.

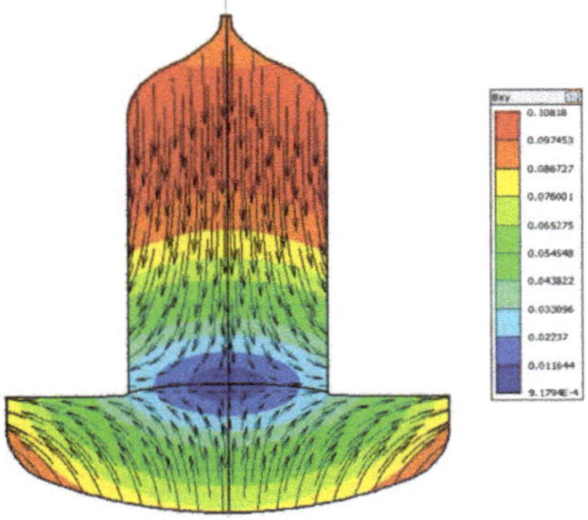

(**a**)

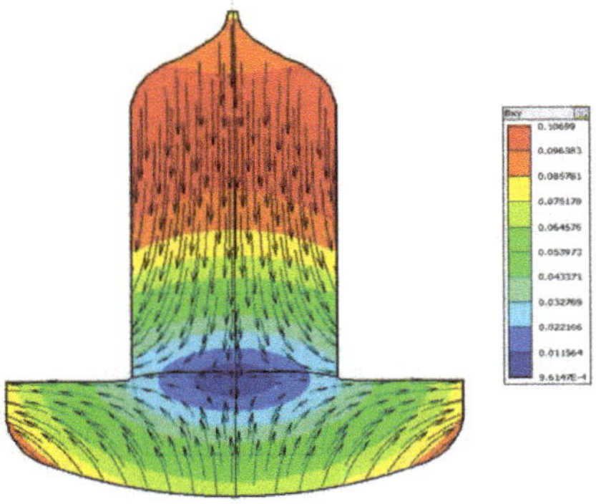

(**b**)

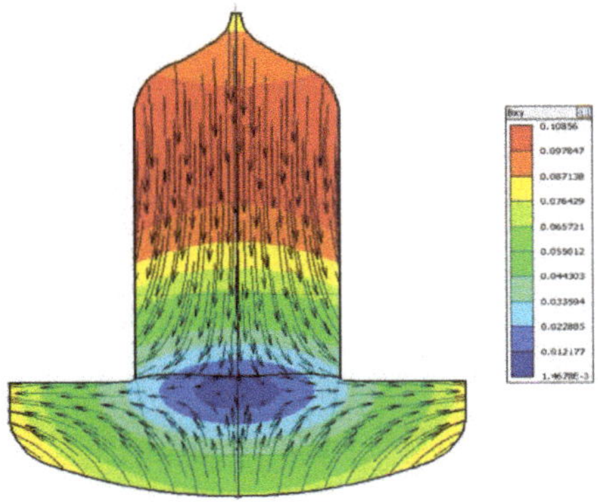

(**c**)

Figure 2. Distribution of B_{xy} with CMF (**a**) 20 mm above the m-g interface, (**b**) on the m-g interface, (**c**) 20 mm below the m-g interface.

In this paper, the initial oxygen concentration behavior in the silicon crystal was studied in three zero-Gauss plane models: ZGP located (a) 20 mm above the melt-gas (m-g) interface, (b) on the m-g interface, and (c) 20 mm below the m-g interface. Previous studies [13] have detailed numerical simulation methods for temperature, melt flow, and oxygen diffusion boundary conditions at the quartz crucible walls. The influence of the magnetic field position on oxygen concentration is based on the diffusion boundary conditions of melt flow velocity at the crucible wall and solid-liquid interface, respectively. All numerical calculations were performed using CGSim (version 23.1) software.

4. Results and Discussion

4.1. CMF Effects on the Silicon Melt Convection

Figure 3 compares the influence of magnetic field intensity and distribution on the flow structure under different CMF conditions. It can be seen from Figure 3 that the structure of the melt under different CMF conditions consists of two parts, one of which forms a Taylor–Proudman unit (1) under the ingot, and the other part forms a buoyancy unit (2) along the crucible wall [14]. Among them, cell (1) is mainly caused by the rotation of the crystal and crucible, and cell (2) is mainly formed by convection driven by buoyancy. Apparently, under the three ZGP models, the distribution area and strength of the two kinds of circulation have obvious changes.It can be noticed from Figure 3 that the Taylor–Proudman vortex (1) is a clockwise flow cycle. As the position of the zero magnetic surface moves down, the horizontal magnetic field near the growth interface gradually increases and dominates because this direction is contrary to the Taylor–Proudman vortex (1), and inhibits the clockwise cell (1) in the crystallization zone. On the contrary, as the zero magnetic surface moves upward, the horizontal magnetic force weakens, and the flow area of the Taylor–Proudman vortex (1) gradually increases and affect the distribution of the flow (2) in the oxygen-enriched parts with melt, thus affecting the oxygen to deliver to the crystallization surface. To more clearly judge the difference in convection strength in the three cases, we intercept the intersection of cells (1) and (2) in Figure 3a–c (shown by the red triangle star in the figure) and compare the flow velocity. The following data are obtained: when the ZGP is +20 mm, the Vx-Vy value is 6.4348×10^{-3} m/s; when the ZGP is 0 mm, the Vx-Vy value is 7.3943×10^{-3} m/s; when the ZGP is -20 mm, the Vx-Vy value is 8.0582×10^{-3} m/s. Therefore, when the zero magnetic plane position is moved down, the flow rate at the intersection increases, so oxygen will have a greater chance of getting into the single crystal.

When the ZGP is at 20 mm above the m-g interface, the CMF suppresses the buoyancy convection near the crucible wall, which is a region of high magnetic field intensity. The cell (1) extends from the central part to the outer region of the melt, and the flow area is the largest, thus preventing the diffusion of oxygen from the bottom of the crucible wall into the solid-liquid interface, as shown in Figure 3a.

In addition, moving the center of the ZGP above the m-g interface increases the vertical component of the CMF in the direction opposite to the buoyancy cell (2). In particular, the ZGP is located 20 mm above the m-g interface, which leads to weak cells (2). Moreover, it can be seen from the magnetic induction intensity distribution in Figure 3c that the bottom of the crucible wall is the weakest in this case. Therefore, the buoyancy cell (2) is enhanced, while the Taylor–Proudman cell (1) is weakened, resulting in a higher oxygen melt at the bottom, entering the crystal. At the same time, the distribution of the two kinds of flow circulations leads to the M-shaped growth interface geometry shown in Figure 3c, which is known as a source generated from dislocation [15] and is not conducive to the single crystal growth process.

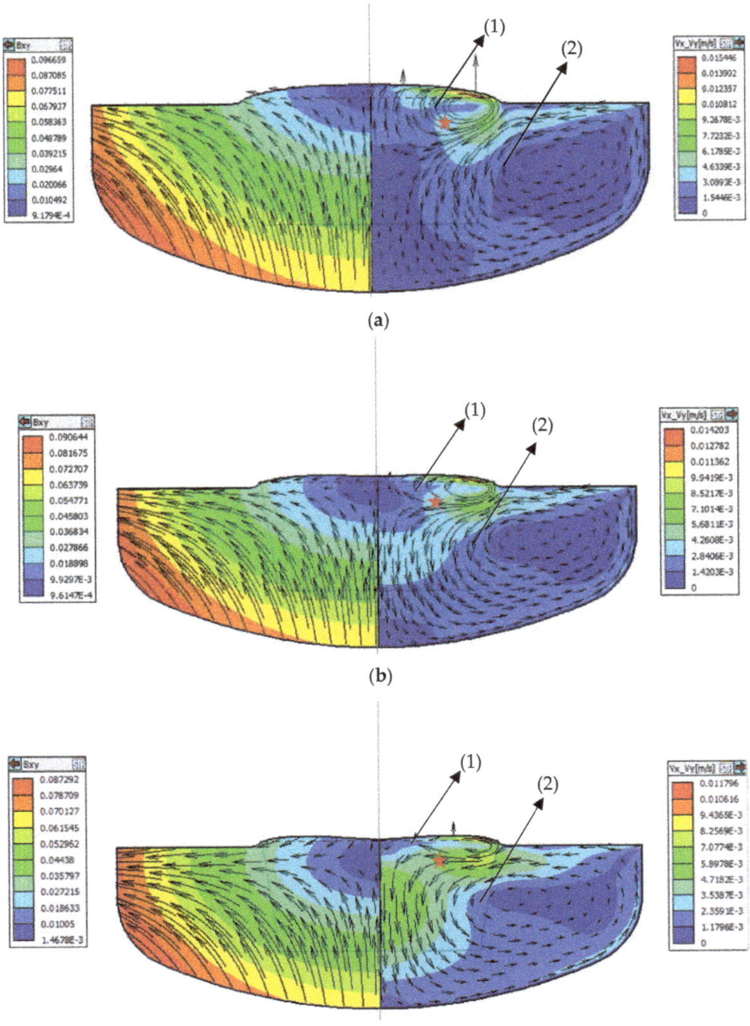

Figure 3. Distribution of magnetic induction in the melt (left) and flow structure (right) with CMF (**a**) 20 mm above the m-g interface, (**b**) on the m-g interface, (**c**) 20 mm below the m-g interface.

4.2. CMF Effects on the Oxygen Transport

The oxygen concentration in the silicon melt is equal to the difference between the dissolution rate of the quartz crucible and the evaporation rate of SiO. It is assumed that oxygen diffusion at all surfaces is of dissolution and evaporation rate determinants, and in the case of steady growth, can establish a simple relationship:

$$[O]_s = [O]_m - [O]_a \tag{1}$$

where, $[O]_S$, $[O]_m$, and $[O]_a$ are the number of oxygen atoms from unit time entering the crystal, melting from the quartz crucible, and evaporating from the free surface of the melt [16–19], respectively.

$$[O]_s = A_R v k_e C_m = A_c D(C_c - C_m)/\delta_c - A_m D(C_a - C_m)/\delta_m \tag{2}$$

where A_R is the cross-sectional area of the crystal rod, A_c is the contact area between the quartz crucible and silicon melt, A_m is the contact area between silicon melt and external atmosphere, v is the pulling speed, k_e is the effective segregation coefficient of oxygen, and C_m is the concentration of oxygen in the melt. C_c is the concentration of oxygen on the surface of the quartz crucible, C_a is the concentration of oxygen in the peripheral atmosphere, δ_c is the thickness of the diffusion boundary layer between the quartz crucible and silicon melt, δ_m is the thickness of the diffusion boundary layer between silicon melt and the peripheral atmosphere, and D is the diffusion coefficient.

The factors that determine the dissolution rate of the quartz crucible are the crucible wall temperature and the contact area between the quartz crucible and the silicon melt. The dissolution rate of the quartz crucible, like most chemical reactions, increases exponentially with an increase in temperature. Therefore, the temperature of the quartz crucible wall will increase the C_c value in Equation (2). Because the center of the melt surface must remain at the freezing point, an increase in the temperature of the quartz crucible wall corresponds to an increase in the temperature gradient inside the silicon melt. In the three models studied in this paper, the weight of the silicon melt is the same as the size of the quartz crucible, so the contact area between the quartz crucible and the silicon melt is fixed, and the area ratio (A_c/A_m) of the silicon melt to the free surface is the same. The dissolution rate of the quartz crucible is mainly affected by temperature, and the temperature distribution under the three models is shown in Figure 4a–c. It can be seen from the figure that the temperature difference between the quartz crucible wall and the temperature difference at the free interface is very small, and the difference is less than 1 k, so there is barely a difference between oxygen dissolution rate and evaporation.

Based on the above discussion, under certain crystal growth parameters, the oxygen concentration in the crystal depends on the thickness of the two diffusion boundary layers. However, convection in the melt will affect the thickness of the diffusion boundary layer, change the transport of oxygen content, and thus affect the distribution of oxygen content, so it is significant to study the convection in the melt. The theoretical calculation of the thickness of the boundary layer is easily derived. In the previous research, the boundary layer at the crucible wall was discussed [20], and there is a substantial theory and calculation method. The habitual force expression of the unit is $\rho v\partial u/\partial x$, and $\partial u/\partial x$ is proportional to U/l (U is the velocity at the outer edge of the boundary layer); therefore $\rho u^2/l$, and the expression of friction force of the monolith volume is $\partial\tau/\partial y$. Suppose that under the condition of stratified flow, its expression is $\mu\partial^2 u/\partial x^2$; since the magnitude of the velocity ladder perpendicular to the wall is equal to the magnitude of the accumulated friction force of the mono site, the lower relation formula can be obtained as follows,

$$\mu U/\delta^2 \sim \rho U^2/l \tag{3}$$

The correlation formula of boundary layer thickness is obtained, below:

$$\delta \sim \sqrt{\mu l/\rho U} = \sqrt{vl/U} \tag{4}$$

The scale factor is 5, taking the length of 1 as the reference length, so it can be calculated as follows,

$$\delta = 5/\sqrt{\mathrm{Re}} \tag{5}$$

The common formula of the deduced boundary layer is as follows,

$$\delta(\mathrm{x}) \propto \sqrt{\mathrm{vx}/\mathrm{U}} \tag{6}$$

It can be concluded from Formula (6) that the boundary layer is inversely proportional to the axial flow velocity, where the thickness of the boundary layer, Re, is the Reynolds number, and $U = V_{(x,y)}$ is the longitudinal velocity.

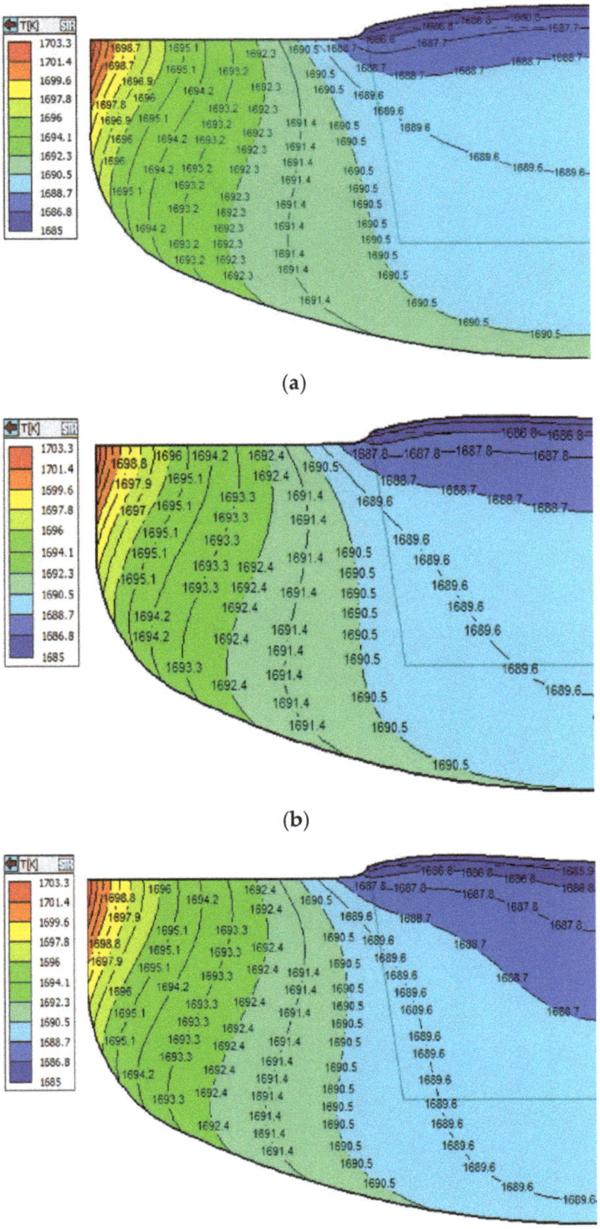

(**a**)

(**b**)

(**c**)

Figure 4. Distribution of temperature in the melt at (**a**) 20 mm above the m-g interface, (**b**) on the m-g interface, (**c**) 20 mm below the m-g interface.

Consequently, to clearly explain the effect on oxygen concentration under the three magnetic field models, we will focus on the convection at the crucible wall, crucible bottom, and free interface. As can be seen from the convection distribution on the right of Figure 3, under the action of buoyancy unit (2), the convection at the crucible wall is dcminated by the longitudinal direction, while the convection at the bottom of the crucible and the free interface is radial and opposite. As can be seen from the description of the first section, the convection under different magnetic field conditions is different. To clearly explain the oxygen concentration under the three magnetic field models, we show a typical example of the distributions of radial velocity components in the melt, as shown in Figure 5a–c.

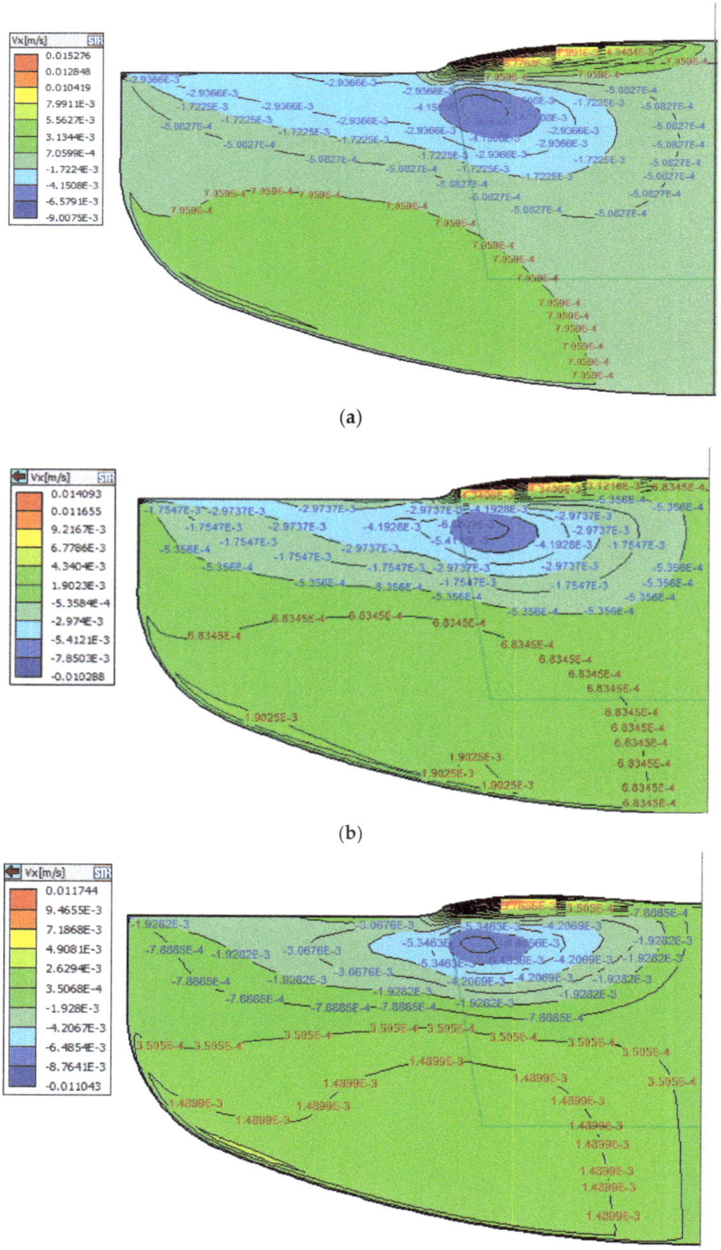

(a)

(b)

(c)

Figure 5. Distribution of radial velocity component (V_x) in the melt at (**a**) 20 mm above the m-g interface, (**b**) on the m-g interface, (**c**) 20 mm below the m-g interface.

It is generally known that the oxygen in the melt comes from oxygen dissolution, which occurs at the interface between the silicon melt and the crucible. The oxygen is mainly dissolved at the bottom of the crucible and then transferred to the vicinity of a solid-liquid interface through thermal convection. According to previous studies and Formula (6), the thickness of the boundary layer is inversely proportional to the velocity at this position. As can be seen in Figure 5a–c, the horizontal velocity at the bottom of the crucible and Angle R gradually increases as the position of the ZGP moves down, and the horizontal convective flow rate of the melt near the bottom of the crucible is the largest at the −20 mm position. Therefore, the diffusion boundary layer between the quartz crucible and the silicon melt at this position is thin, the washing of the crucible wall is more intense, resulting in more oxygen impurities entering the melt, and finally, a part of the oxygen in the melt is integrated into the crystal. The oxygen concentration distribution in the three cases is displayed in Figure 6a–c. Through the comparison of the results of oxygen concentration, the oxygen concentration near the bottom of the crucible gradually increases as the position of the ZGP moves down, and it is again verified that the decomposition of the crucible wall is inversely proportional related to the strength of the convection flow (that is the thickness of the boundary layer).

The influence of different CMF configurations on the flow pattern has been studied above, and the initial oxygen concentration distribution in the three cases is displayed in Figure 6a–c. In the CMF, most of the oxygen is carried to the free surface and evaporates. Only a small amount is transported to the crystallization zone. A small gradient of oxygen concentration is found in the core region of the melt due to the Taylor–Proudman flow.

When the ZGP is located 20 mm above the m-g interface, according to the above discussion and V_X-V_y overall analysis, the flow rate near the crucible wall is slow due to the inhibition of the melt convection near the crucible wall, which results in a decrease in the dissolved oxygen concentration in the crucible wall and the slow corrosion rate of the crucible. Moreover, with the melt convection of cell (2), oxygen in the area near the crucible wall is transported to the free interface and volatilized. Less oxygen is delivered to the core region of the melt, and less and less oxygen diffuses into a crystal, so the low-oxygen zone below the solid-liquid interface is larger, as shown in Figure 6a.

When the ZGP is located on the m-g interface at a given magnetic field strength, the zero magnetic surface enhances the flow of dissolved oxygen from the crucible into the melt, then brings it into the crystal. However, a decrease in the axial magnetic intensity in the crystallization zone causes the radial distribution of oxygen on the growth interface to become uniform.

When the ZGP is located 20 mm below the m-g interface cases, with the change in the position of the zero magnetic surface, the strength and weakness of the convection flow near the crucible wall changed with it. Under the action of strong convection, the melt convection brushing near the wall of the quartz crucible is stronger. This can also be proved by the melt convection intensity distribution oxygen concentration at the growth interface, which increases, greater than at 20 mm above the m-g interface.

In the above-mentioned, single-crystal growth parameters (experiments are carried out with fixed rotation rates for the crucible and crystal), we carried out the actual industrial single-crystal pulling process. The whole single crystal is segmented, and a slug is taken at the segmented position and tested by Fourier transform infrared spectroscopy (FT-IR) to obtain the central oxygen concentration of the head and tail of each crystal segment. The results are shown in Figure 7. The figure displays the difference in oxygen concentration in the center of the interface at the crystal segment with the crystal length. When the ZGP is located 20 mm above the m-g interface, the overall oxygen concentration shows the lowest level. At the former part of crystal lengths, the axial oxygen concentration decreases slowly as the crystal length increases until it reaches a minimum; the lowest value in the three groups was 1.02 ppma. Then it increases continuously as the crystal length increases further. Therefore, the largest oxygen concentration appears in the initial and final stages of the whole crystal growth process. The experimental results and simulation results show the same trend, which further confirms the discussion under numerical simulation in this paper.

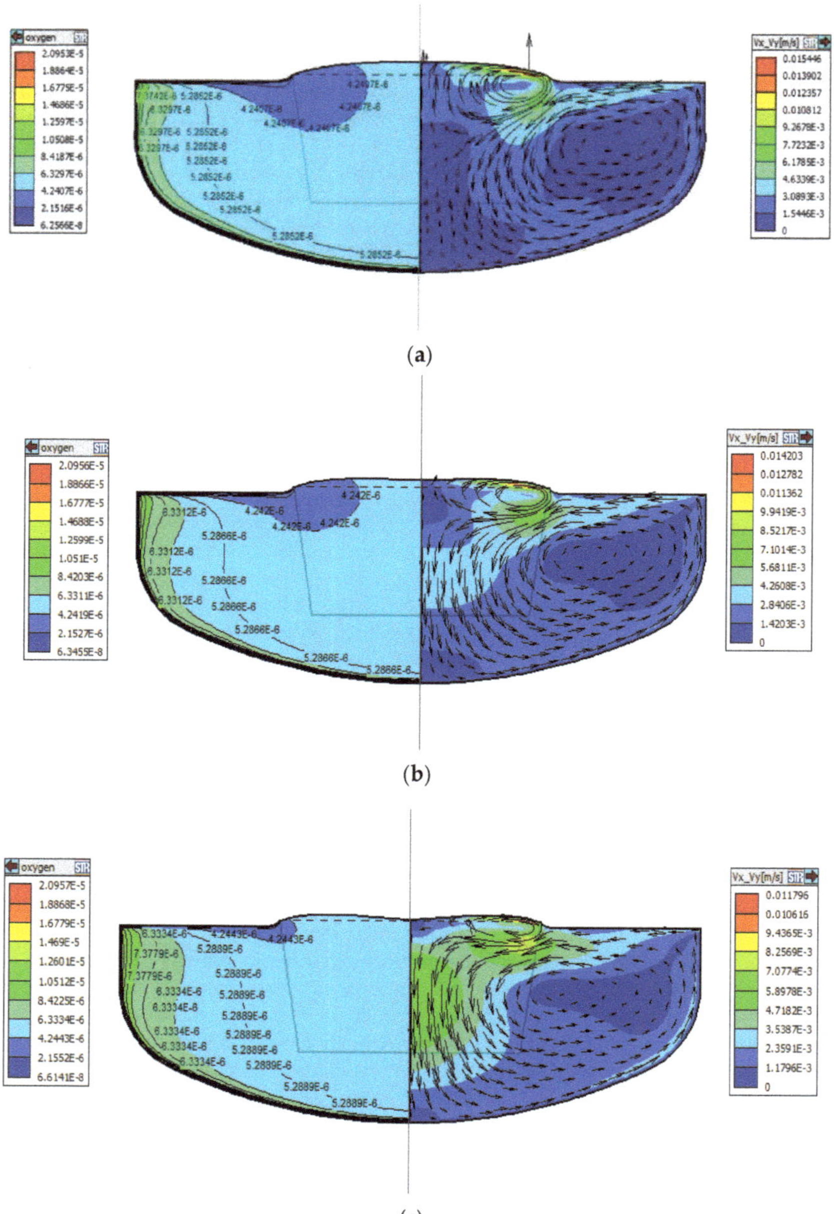

Figure 6. Distribution of oxygen concentration in the melt (left) and flow pattern (right) with CMF (**a**) 20 mm above the m-g interface, (**b**) on the m-g interface, (**c**) 20 mm below the m-g interface.

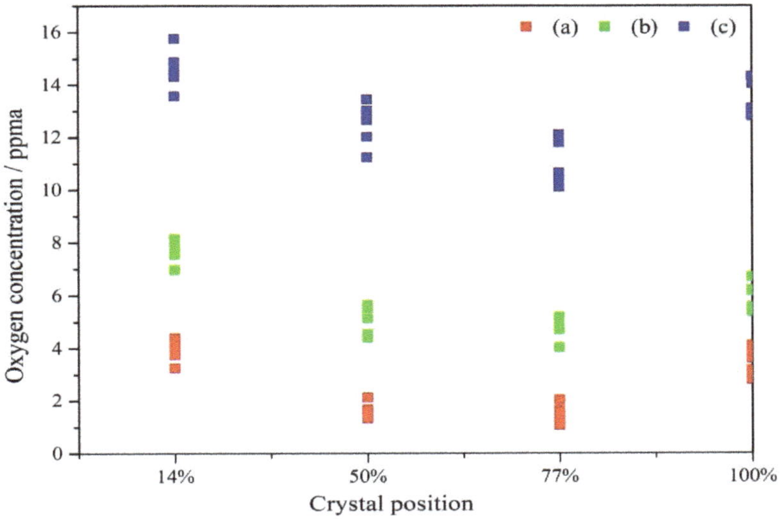

Figure 7. Experimental data of oxygen concentration with CMF (a) 20 mm above the m-g interface, (b) on the m-g interface, (c) 20 mm below the m-g interface.

4.3. CMF Effects on Growth Interface

We compared the shape of the growth interface under different CMFs; the simulation results are shown in Figure 8. When the ZGP is located 20 mm above the m-g interface, the interface is convex to the crystal. While the ZGP is located on the m-g interface, the interface near the crystal axis is slightly concave to the crystal and convex at the right side near the edge of the crystal. When the ZGP is located 20 mm below the m-g interface, the growth interface becomes more convex towards the liquid direction, and a W-shaped growth interface is obtained. The interface shape is affected by the convection of the Taylor–Proudman unit (1) near the growth interface, as shown in Figure 3. In addition, it can be seen that the lower the position of the ZGP, the smaller the temperature gradient below the crystal, so the more convex near the crystal axis.

To confirm the morphology of the growth interface in the actual crystal growth and verify the guiding significance of the numerical simulation, we tested the growth interface of the single crystal in the actual industrial single-crystal pulling process. The results are shown in Figure 9. The red line is the location of the single crystal length discussed in the simulation. The experiment uses the LPCon-LP SCAN (Lateral Photovoltage Scanning) tool to select single crystal profile samples with a range of 50 mm length above and below the simulated single crystal position for longitudinal scanning. The sample is coated with gallium indium alloy to make them conductive, and the image of solid-liquid interface streaks of the profile is shown by infrared ray scanning through the sample. The results show that the growth interface on the surface under the three ZGP positions is consistent with the simulation trend, which further confirms that the actual production has a positive match with the numerical simulation.

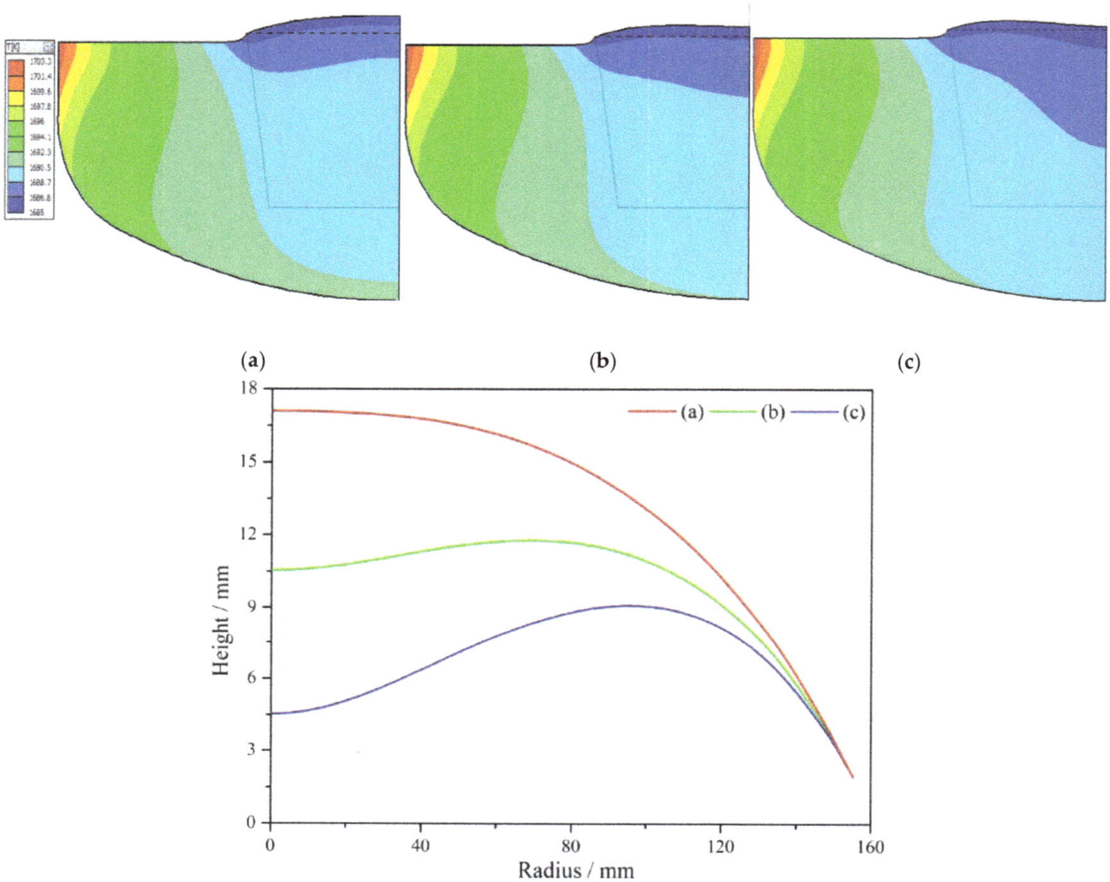

Figure 8. Comparison between growth interface shapes under different CMFs in simulation (**a**) 20 mm above the m-g interface, (**b**) on the m-g interface, (**c**) 20 mm below the m-g interface.

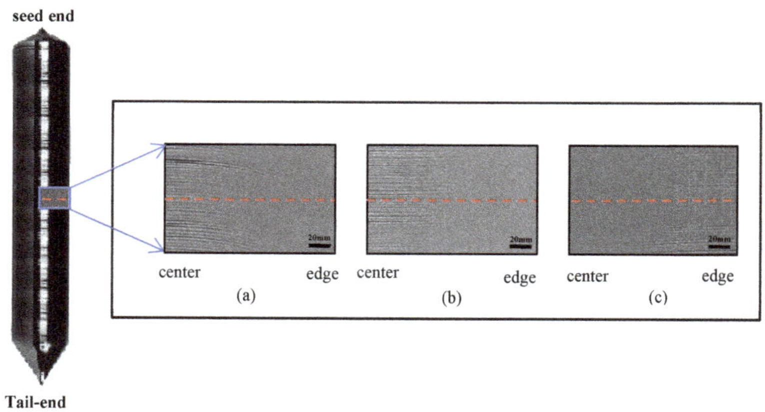

Figure 9. Comparison between growth interface shapes under different CMFs in experiments (**a**) 20 mm above the m-g interface, (**b**) on the m-g interface, (**c**) 20 mm below the m-g interface.

5. Conclusions

This paper investigated the effects of CMF distribution at different positions of the zero-Gauss plane (ZGP) on melt convection, oxygen concentration transfer, and the crystal-melt interface of a 300 mm MCZ silicon crystal by conducting numerical simulations. The results show that when the ZGP moves downward gradually from the melt-gas interface, the melt flow velocity near the quartz crucible wall gradually increases, resulting in a decrease in the thickness of the diffusion boundary layer. Along with the changes in the convection area and intensity, more oxygen is generated and transported to the bottom of the solid-liquid interface, and eventually diffuses into the crystal, leading to a significant increase in oxygen concentration. In the oxygen concentration distribution of the crystal discussed in this paper, the ZGP located at +20 mm case is the lowest value in the calculation. In addition, the melt convection and temperature gradient at the solid-liquid interface lead to different shapes of the growth interface. When the ZGP is at +20 mm, the shape of the interface is convex towards the crystal, while at −20 mm, the interface near the crystal axis is slightly concave to the crystal. Additionally, the actual industrial crystal growth can be accomplished by simulation results, and the crystal data were evaluated. The trend of crystal oxygen concentration is consistent with the simulation analyses, and the scanning results of a solid-liquid interface are in line with the simulation results as well, indicating that the simulation and discussion in this paper provide valuable guidance for practical production.

Author Contributions: Formal analysis, H.C., H.L. and G.C.; Data curation, H.C.; Writing—original draft, C.S., X.A., H.Z. and G.C.; Writing—review & editing, H.Z., H.L. and G.C. All authors have read and agreed to the published version of the manuscript.

Funding: This work was supported by the project for the Science and Technology Correspondent of Tianjin City (Grant No. 20YDTPJC01710) and the Research Foundation of Education Bureau of Hebei (Grant No. QN2021044). Also Special thanks to Miao Wang and Mingchao Hou for their contributions in the experiments and Fab resource allocation.

Institutional Review Board Statement: Not applicable.

Informed Consent Statement: Not applicable.

Data Availability Statement: Not applicable.

Conflicts of Interest: The authors declare no conflict of interest.

References

1. Schulze, H.-J.; Ofner, H.; Niedernostheide, F.-J.; Lukermann, F.; Schulz, A. Fabrication of IGBTs using 300 mm magnetic Czochralski substrates. *IET Power Electron.* **2019**, *12*, 3870–3873. [CrossRef]
2. Kalaev, V.; Sattler, A.; Kadinski, L. Crystal twisting in Cz Si growth. *J. Cryst. Growth* **2015**, *413*, 12–16. [CrossRef]
3. Bok-Cheol, S.; In-Kyoo, L.; Kwang-Hun, K.; Hong-Woo, L. Oxygen concentration in the Czochralski-grown crystals with CUSP-magnetic field. *J. Cryst. Growth* **2005**, *275*, 455–459.
4. Liu, X.; Liu, L.; Li, Z.; Wang, Y. Effects of CUSP-shaped magnetic field on melt convection and oxygen transport in an industrial CZ-Si crystal growth. *J. Cryst. Growth* **2012**, *354*, 101–108. [CrossRef]
5. Kakimoto, K.; Liu, X.; Nakano, S. Analysis of the Effect of CUSP-Shaped Magnetic Fields on Heat, Mass, and Oxygen Transfer Using a Coupled 2D/3D Global Model. *Cryst. Res. Technol.* **2022**, *57*, 2100092. [CrossRef]
6. Koichi, K.; Kyung-Woo, Y.; Minoru, E. Oxygen transfer during single silicon crystal growth in Czochralski system with vertical magnetic fields. *J. Cryst. Growth* **1996**, *163*, 238–242.
7. Liu, X.; Liu, L.; Li, Z.; Wang, Y. Effects of static magnetic fields on thermal fluctuations in the melt of industrial CZ-Si crystal growth. *J. Cryst. Growth* **2012**, *360*, 38–42. [CrossRef]
8. Borisov, D.; Artemyev, V.; Kalaev, V.; Smirnov, A.; Kuliev, A.; Zobel, F.; Kunert, R.; Turan, R.; Aydin, O.; Kabacelik, I. Advanced approach for oxygen transport and crystallization front calculation in Cz silicon crystal growth. *J. Crystal Growth* **2022**, *583*, 126493. [CrossRef]
9. Nguyen, T.-H.; Chen, J.-C.; Hu, C.; Chen, C.-H. Effects of crystal-crucible iso-rotation and a balanced/unbalanced cusp magnetic field on the heat, flow, and oxygen transport in a Czochralski silicon melt. *J. Cryst. Growth* **2020**, *531*, 125373. [CrossRef]
10. Gunjal, P.R.; Ramachandran, P.A. Tailoring the oxygen distribution in 300 mm Czochralski crystal of pure silicon using cusp magnetic field. *Prog. Comput. Fluid Dyn. Int. J.* **2010**, *10*, 307–318. [CrossRef]

11. Eglitis, R.I.; Purans, J.; Popov, A.I.; Jia, R. Tendencies in ABO_3 perovskite and SrF_2, BaF_2 and CaF_2 bulk and surface F-center Ab initio computations at high symmetry cubic structure. *Symmetry* **2021**, *13*, 1920. [CrossRef]
12. Ding, J.; Li, Y.; Liu, L. Effect of CUSP magnetic field on the turbulent melt flow and crystal melt interface during large-size Czochralski silicon crystal growth. *Int. J. Therm. Sci.* **2021**, *170*, 107137. [CrossRef]
13. Chen, J.C.; Guo, P.C.; Chang, C.H.; Teng, Y.Y.; Hsu, C.; Wang, H.M.; Liu, C.C. Numerical simulation of oxygen transport during the Czochralski silicon crystal growth with a CUSP magnetic field. *J. Cryst. Growth* **2014**, *401*, 888–894. [CrossRef]
14. Raufeisen, A.; Breuer, M.; Botsch, T.; Delgado, A. Prediction of the three-phase boundary movement in Czochralski crystal growth. *J. Cryst. Growth* **2010**, *312*, 2297–2299. [CrossRef]
15. Smirnova, O.V.; Durnev, N.V.; Shandrakova, K.E.; Mizitov, E.L.; Soklakov, V.D. Optimization of furnace design and growth parameters for Si Cz growth, using numerical simulation. *J. Cryst. Growth* **2008**, *310*, 2185–2191. [CrossRef]
16. Murgai, A. Mechanisms of oxygen incorporation in Czochralski-grown silicon. In *Semiconductor Silicon 1981 (Proceedings of the 4th International Symposium on Silicon Materials Science and Technology, Electrochemical Society PV 81-5)*; Huff, Y.T.H., Kriegler, R., Eds.; The Electrochemical Society: Pennington, NJ, USA, 1981; p. 11.
17. Hoshikawa, K.; Hirata, H.; Nakanishi, H.; Ikuta, K. Control of oxygen concentration in Cz silicon growth. In *Semiconductor Silicon 1981 (Proceedings of the 4th International Symposium on Silicon Materials Science and Technology, Electrochemical Society PV 81-5)*; Huff, Y.T.H., Kriegler, R., Eds.; The Electrochemical Society: Pennington, NJ, USA, 1981; p. 101.
18. Carlberf, T.; King, T.B.; Witt, A.F. Dynamic Oxygen Equilibrium in Silicon Melts during Crystal Growth by the Czochralski Technique. *J. Electrochem. Soc.* **1982**, *129*, 189. [CrossRef]
19. Moody, J.W. Oxygen precipitation in Czochralski silicon. In *Semiconductor Silicon 1981 (Proceedings of the 4th International Symposium on Silicon Materials Science and Technology, Electrochemical Society PV 81-5)*; Huff, Y.T.H., Kriegler, R., Eds.; The Electrochemical Society: Pennington, NJ, USA, 1981; p. 100.
20. Brice, J.C.; Laudise, R.A. *The Growth of Crystals from Liquids*; North-Holland Publishing Company: Amsterdam, The Netherlands, 1973; Chapter 4.

coatings

Article

Simulation of the Inductor Structure to Improve FZ Thermal Fields

Xingtian Ai [1,2], Chenguang Sun [1,2], Hui Zhang [1,3,*], Jian Sun [2], Luxiao Xie [2], Guodong Liu [1,3,*] and Guifeng Chen [1,3,*]

1 School of Materials Science and Engineering, Hebei University of Technology, Tianjin 300132, China; 13363361297@163.com (X.A.); sunchenguang@zhlxsemicon.com (C.S.)
2 Zhonghuan Advanced Semiconductor Materials Co., Ltd., Yixing 214200, China; sunjian@zhlxsemicon.com (J.S.); xieluxiao@zhlxsemicon.com (L.X.)
3 Hebei Engineering Laboratory of Photoelectronic Functional Crystals, Hebei University of Technology, Tianjin 300130, China
* Correspondence: zhanghui_1207@163.com (H.Z.); gdliu1978@126.com (G.L.); cgfchen@hebut.edu.cn (G.C.)

Abstract: The floating zone (FZ) is one of the important methods for pulling silicon single crystals, but there are still problems of an unstable thermal field and crystallization difficulties. They will directly affect the growth of single crystals, resulting in defects and even fractures, seriously reducing production efficiency. Based on this, the effect of the modified inductor structure on the FZ thermal field is investigated in this paper. Using COMSOL 6.0 simulation software, 2D and 3D FZ models are established. The inductor steps under the 2D model and the inductor slits under the 3D model are compared to analyze the effects of steps and slits on the 8-inch FZ thermal field and melt flow. The distributions of temperature fields and melt flow in the melting zone under the action of the axial magnetic field are calculated by finite element analysis. The results show that the melt under the introduction of steps in the 2D model and the cross-slit structure in the 3D model is the most stable and favorable for crystal growth, which matches the actual production.

Keywords: floating zone; inductor structure; thermal field

1. Introduction

Zone melting is a crucial method for semiconductor material purification and silicon single-crystal growth because the melting zone is suspended between the polysilicon rods and the silicon single crystal; it is also known as the suspended floating zone (FZ) melting method [1], as shown in Figure 1a. The FZ method uses a high-frequency induction principle and skin collection effect through the coil to realize the local surface heating of polysilicon rods, with the help of the surface tension of the molten silicon and the electromagnetic torus buoyancy. The melting zone is in a state of suspension and then in the specific crystal direction of the seed crystal guidance. The melting zone solidification generates a single crystal. The current FZ technique is a modification of the previous technique, where the inductor hole is enormous, and the melt has no "waist", which causes the single crystal to grow unstably due to gravity, resulting in defects and fracture [2]. The current FZ technique, also known as the needle-eye technique [3], has a very narrow hole in the center of the inductor to form a skinny "waist" and a small melt [4], as shown in Figure 1b. The principle is to use a high-frequency induction inductor as the heat source, with the feed rod and single crystal above and below the inductor, respectively, and the resulting melt is suspended above the inductor and below the feed rod. The melt is suspended above the inductor and below the feed rod. Compared to the Czochralski (CZ) technique [5], the FZ technique does not require a crucible [6]. The single crystals have shallow oxygen content and low dislocation density [7] and are characterized by high purity and quality, resulting in a significant reduction of impurities in silicon and a high minority carrier lifetime but also higher cost. The semiconductor industry has multiplied in recent years, and the purity of CZ silicon has failed to meet the standards of some materials manufacturing. The production of 8-inch FZ silicon is the current mainstream direction, and FZ silicon single crystals can be used in IGBT, MEMS, high-efficiency solar energy, and image sensors. With the development of precision device technology, reducing costs and producing large-size, high-quality FZ silicon single crystals is imperative.

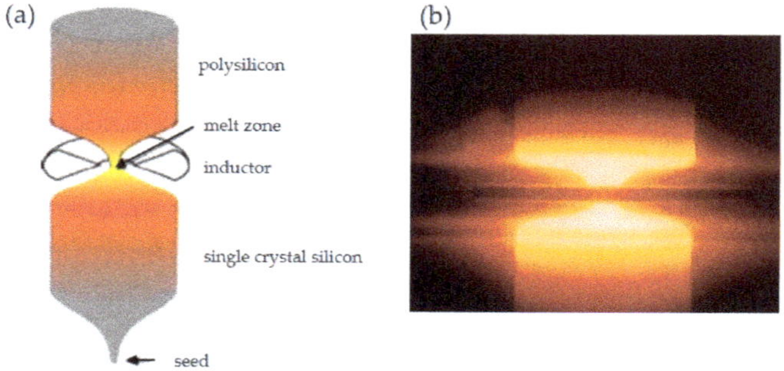

Figure 1. The FZ process (**a**); Schematic diagram of the melt section (**b**). Reprinted with permission from [1]; Copyright 2018 Elsevier.

The FZ technique uses the principle of induction heating [8]; the frequency is generally 2–3 MHz [9], and the high voltage DC input is through the power supply cabinet to the oscillating circuit inside the outer groove of the FZ furnace. Finally, the output high-frequency alternating current will produce an electromagnetic (EM) field near the induction inductor, which is essentially a wave function that changes with time. The feed rod part near the inductor will produce eddy currents, and due to its high purity and high resistance, there will be a large amount of Joule heat, which is induction heating. In this paper, based on the numerical simulation study of the EM field, temperature field, and flow field, the structure of the inductor is improved to improve the thermal field [10].

This calculation was performed using COMSOL large simulation software, which has the advantage of wide applicability. Its calculation principle, in which each introduced physical field has its partial differential equation solution formula, was calculated by establishing the geometry, defining the material, introducing multiple physical fields, setting boundary conditions, dividing the mesh, and coupling multiple physical fields [11].

Before the calculation, we need to analyze the relationship between each physical field of the FZ process [12]. Figure 2 shows that specific process parameters and inductor structures affect the physical fields, and the physical fields are coupled with each other [13]. It can also be seen from the figure that the inductor structure dramatically influences the EM field; this is because changes in the inductor structure affect the distribution of the EM field. A change in the EM field causes a difference in the distribution of force and Joule heat in the melt; it further influences the temperature field and flow field. Based on this, it is important to improve the inductor structure.

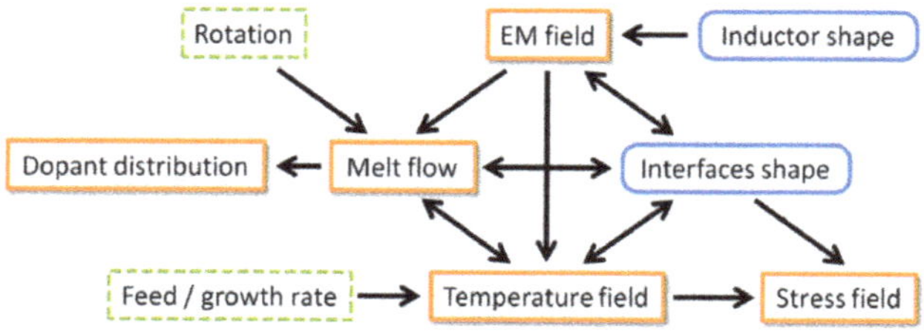

Figure 2. Coupling relationship between model physical fields. Reprinted with permission from [12]; Copyright 2015 Elsevier.

For the problems of crystallization difficulties and poor crystal quality in the actual production, in this study, we will improve the temperature field and flow field in the melt, design two kinds of inductors in 2D and 3D, and establish an FZ model to analyze them and come up with a suitable and effective inductor structure to improve the production efficiency.

2. Numerical Model and Calculation Method

We created 2D and 3D models and imported them into COMSOL Multiphysics software. COMSOL is a large-scale advanced numerical simulation software, formerly a MATLAB toolbox [14], and later developed into an independent commercial software. COMSOL is based on the finite element method, which is very suitable for solving partial differential equations and systems of equations. It is rich in preconfigured specialized modules covering various disciplines such as electrical, structural mechanics, fluids, heat transfer, chemical engineering, etc., which can simulate and model any system based on physical fields. It can be used to model and simulate any physical field-based system, providing a graphical user interface (GUI), predefined user interfaces, and associated simulation and analysis tools for typical simulation applications. A wide range of specialized tool modules enable numerical computation in specific research directions. At the same time, a rich set of interfaces allows easy interaction with a wide range of third-party software. COMSOL is a general-purpose software platform for modeling and simulating multi-physical-field problems based on advanced numerical computation methods [15]. There are more than twenty physical field modules. In addition, we can couple any two related physical modules. If the pre-built modules do not meet the requirements, we can also define our partial differential equations (sets) to construct models to solve the problems. There are dozens of sub-modules for users to choose from. In addition, users can also be connected to the software platform through electromagnetism, thermodynamics, and other fields of specialized physical interfaces [16], further expanding the platform's functionality. The advantages of COMSOL for FZ process simulation are the software's ability to perform multiphysics field coupling, its good suitability for this study, and the integration of modeling, computation, and post-processing for easy data handling. Figure 3 introduces the basic modules of COMSOL Multiphysics.

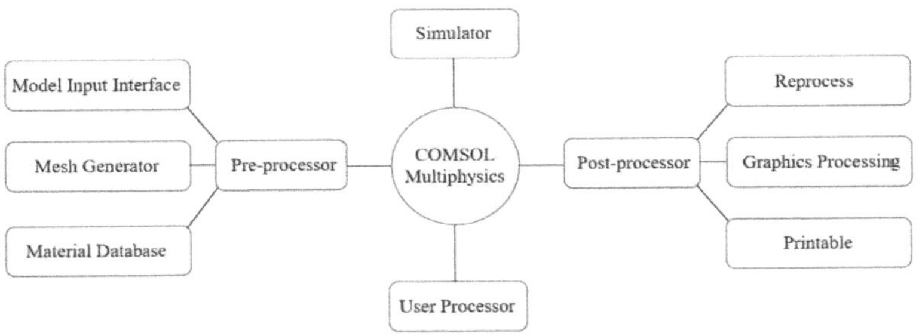

Figure 3. Module introduction of COMSOL.

Initially, the inductors used in the preparation of single crystals by the FZ method had no steps and secondary slits, only a main slit and a central hole, while later, to better melt polysilicon as well as to adapt to the trajectory of polysilicon melting, beveled surfaces were introduced on the surface of the inductors, and based on the basic understanding of the inductor structure. In the present study, it has been improved.

To compare the effects of inductor steps and slits on the melt, 2D inductors were designed with and without steps, as shown in Figure 4; 3D inductors were designed with one and cross shapes for comparison, and the models are shown in Figure 5. After importing the model to define the material and add physical fields, including a magnetic field (mf), solid and fluid heat transfer (ht), and laminar flow (spf) and after coupling electromagnetic heat (emh), non-isothermal flow (nitf), and Marangoni effect (mar), under the magnetic field module, following Ampere's law, the relative permeability is chosen for the magnetization model and calculated as follows:

$$B = \mu_0 \mu_r H \quad (1)$$

where B is the magnetic induction strength, and H is the magnetic field strength. Assuming that the current flows on the surface [17], the approximation considers the feed rod, melt, and single crystal to be axisymmetric and incorporates rotation under the 3D model, which is more conducive to analyzing the effects of changes in inductor structure, under the heat transfer module, setting the boundary temperature, and the crystallization and melting interface is set to the melting point temperature. This module is coupled with laminar flow, and only the effect of buoyancy is considered due to the slight change in material density. It is easier to converge using the Boussinesq approximation to calculate the non-isothermal flow. Due to the presence of tension gradients (caused by temperature gradients) on the surface of the melt, it is also necessary to include the Marangoni effect for the analysis.

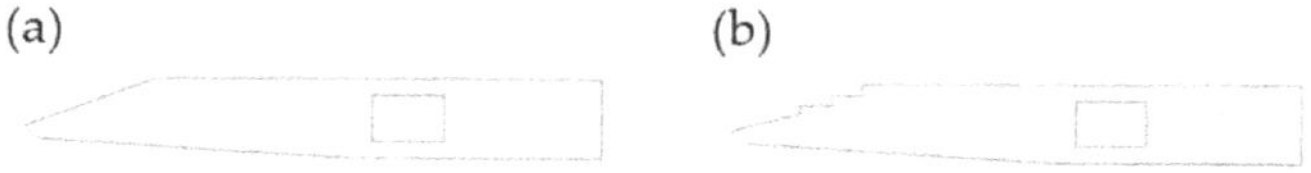

Figure 4. Inductors without steps (**a**) and with steps (**b**).

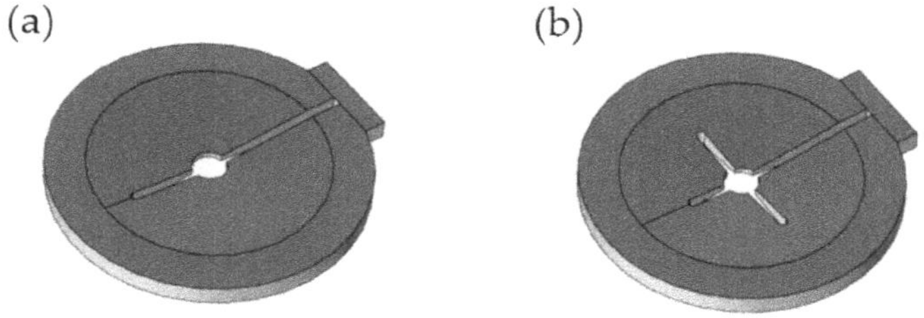

Figure 5. One-way inductor (**a**); cross-way inductor (**b**).

When setting up the grid, due to the complexity of the model, a different grid division is required for each region, and a user-controlled grid, generally a free tetrahedral grid [18], is chosen to refine the grid for the 2D and 3D melt parts, respectively, as shown in Figures 6 and 7. Figure 6 refines the melt mesh and shows the boundary layer refinement due to the apparent solute boundary layer at the growth interface and the powerful free surface interactions (EM forces, Marangoni forces, etc., which will be described below) resulting from flow, so a finer grid is used for the calculation. Figure 7a shows the portion of the melt obscured by the inductor, and we have enlarged the mesh in the part of the melt so that we can better observe it. The finer the grid, the more accurate the calculation. A schematic diagram of the overall structure of the FZ process is given in Figure 7b.

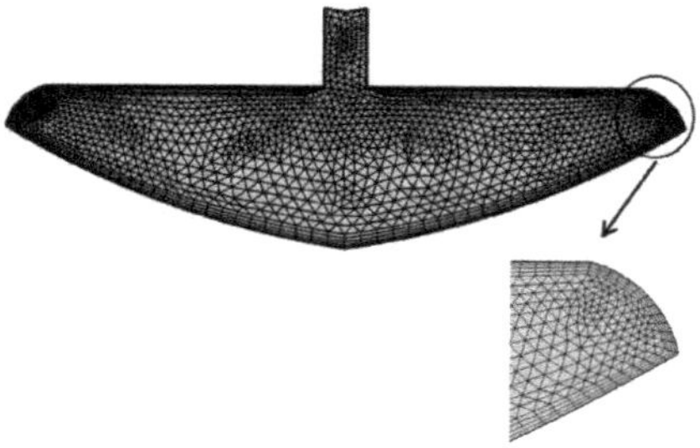

Figure 6. Overall mesh of 2D melt and a localized zoom mesh for the boundaries.

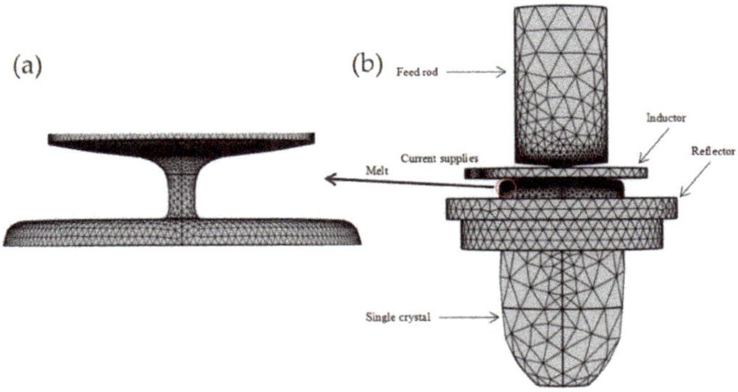

Figure 7. Enlarged 3D melt meshing (**a**); overall FZ meshing (**b**).

3. Introduction to Physical Fields

3.1. The Electromagnetic Field

The principle of induction heating mentioned is to generate an EM field around the inductor using high-frequency currents and produce a large amount of Joule heat. The calculation of alternating currents at high frequencies is based on the time-harmonic assumption to set the equation:

$$\nabla \times \left(\mu_0^{-1} \nabla \times A\right) + \sigma \nabla V + j\omega\sigma A = J_e \quad (2)$$

where A is the magnetic induction, μ_0 is the permeability, σ is the conductivity, V is the potential, j is the imaginary part unit, ω is the angular frequency, and J_e is the generated current density.

In the early days, modeling was done using a 2D model, which saved computational time. In the 2D results, the contour line distribution of the magnetic vector potential is shown, as shown in Figure 8a. Due to the simple two-dimensional model, the central magnetic field is reduced. The external electromagnetic field of the silicon is enhanced, generating a high-temperature region at the free surface of the melt, as shown in Figure 8b, which makes the 2D and 3D results fit better [19]. A part of the line is also present inside the inductor, but practically all the line distribution is present between the inductor and the silicon surface [20].

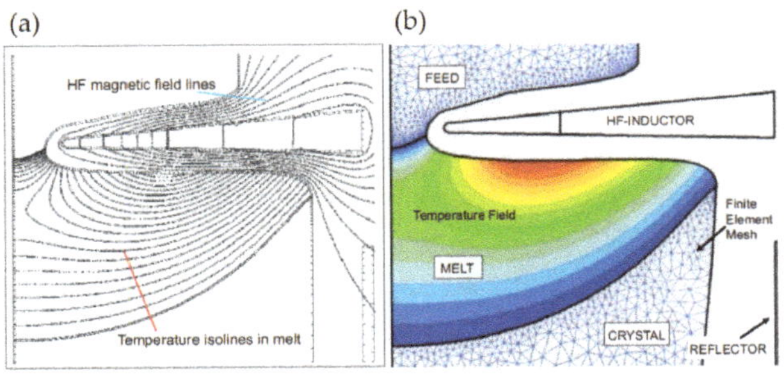

Figure 8. 2D finite element mesh in the solid silicon and the distribution of the temperature field in the melt (**a**); Corresponding high-frequency magnetic field lines and isothermal distributions (**b**).

Figure 2 also shows that the EM field and the interface shape are strongly coupled, based on the fact that the surfaces are axisymmetric, but the inductor is asymmetric to the melt during the FZ process. The simulation process requires defining the EM heat source and calculating the EM force. The 3D model is divided into a triangular mesh, as shown in Figure 9a. Based on the skin effect of the current, the current is mainly concentrated on the inductor surface with high surface current density, which contributes to a feed rod. The current flow differs on the silicon surface and in the inductor slit. On the silicon surface, the current flows almost parallel to the inductor. When the current passes through the inductor slit region, it flows vertically upward to the feed rod and single crystal, respectively, so that the density of the EM heat source is minimized in the slit region, as shown in Figure 9b [21]. Therefore, optimizing the inductor slit design can change the current distribution, melt temperature distribution, and flow.

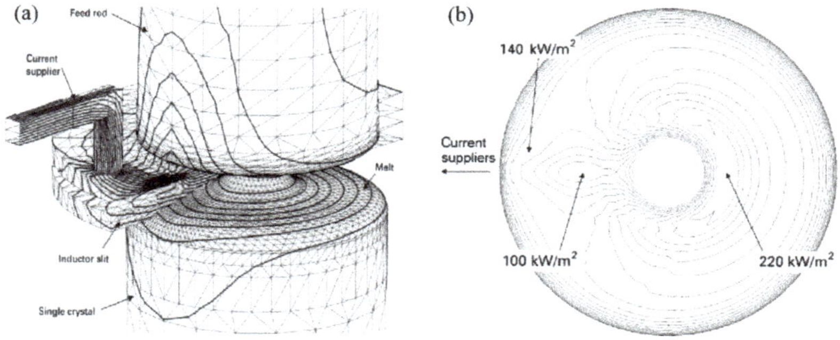

Figure 9. Simulation results of the 3D EM field, showing the EM model of the FZ growth system with only half of the inductors (**a**); Power density at the free surface of the melt (**b**). Reprinted with permission from [12]; Copyright 2015 Elsevier.

3.2. Melt Temperature and Flow

The EM field distribution determines the melt flow, which is closely related to the temperature distribution, and its temperature distribution should be studied first before studying the flow. Figure 10 shows the temperature distribution of the melt in three dimensions as well as the temperature distribution of the cross-section [22], where a high-temperature gradient is formed in the melt by direct heating of the induction inductor and shows high current density and temperature at the side silt, while the main silt has low current density and temperature, showing an asymmetric temperature field, the temperature field is calculated as follows:

$$\rho U \nabla \cdot (h) = \nabla \cdot (a \nabla h) \tag{3}$$

where h is the enthalpy, and a is the diffusivity. The latent heat is calculated as follows:

$$Q_L = \rho V_g L \tag{4}$$

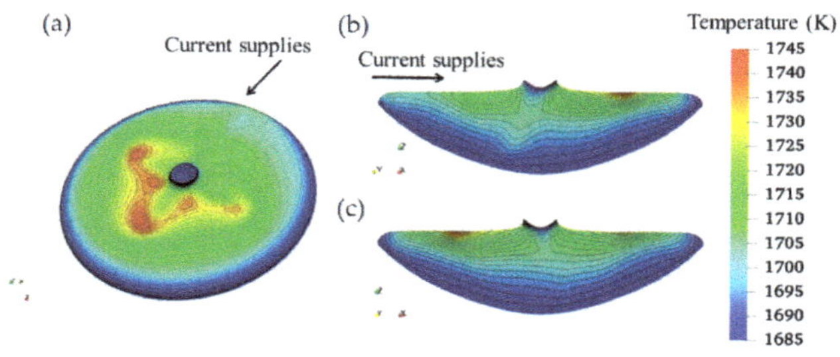

Figure 10. Temperature distribution at the free surface of the melt (**a**); Temperature distribution of the cross-section parallel to the current (**b**); Temperature distribution of the cross-section perpendicular to the current direction (**c**). Reprinted with permission from [21]; Copyright 2020 Elsevier.

V_g is the local crystal growth rate, and L is the latent heat of silicon. The surface current density distribution in the EM model is coupled to the heat production at the free surface of the melt in the heat transfer model by the following equation:

$$Q_{EM} = \frac{J_s^2}{\delta_Q} \tag{5}$$

Q_{EM} and J_s are the surface power density and the current density at the free surface of the melt. The temperature distribution in the melt is made as uniform as possible in a feed rod and single crystal rotating environment. Figure 11 shows the action of different forces in the melt [23]. Due to the weak penetration of the *EM* field in the melt part, a transverse shear stress is formed at the free surface boundary of the melt, and this force contains the *EM* and Marangoni forces [24], and the melt also contains the buoyancy force, which acts on the whole melt. For the temperature field, boundary conditions need to be set, with the melting boundary and the interaction of *EM* and Marangoni forces as the first and second type of boundary conditions, respectively:

$$F = F_{EM} + F_{Ma} = \frac{1}{4}\mu_0\delta\nabla_s(J_s^2) + \frac{\partial\gamma}{\partial T}\nabla_s(T) \tag{6}$$

where γ is the surface tension, the Marangoni force is due to the existence of the surface tension gradient. The direct influence on the surface tension gradient in a melt is the temperature gradient. The temperature at the melting and crystallization interface is low, making the Marangoni force to the two ends of the diffusion, which is unfavorable for the melt. The direction of EM force is opposite to it, which requires the control of the magnetic field to weaken the Marangoni effect. Since the single crystal and feed rod are rotating during the FZ process, it makes the angular velocity component in the melt, which is strongly coupled with the warp velocity itself. It affects the melt flow together, while the rotation also inhibits the flow in the warp direction. It can be seen from the figure that the EM, Marangoni, and buoyancy forces form a closed loop in the melt region, and the presence of these forces causes the existence of vortices in the melt. Vortex 3 is caused by the EM force due to the dense magnetic induction lines near it; vortex 1 is caused by the Marangoni and buoyancy forces and is unstable; and vortex 2 is driven by the buoyancy force alone, and the generation of these vortices affects the distribution of RRV, so that when pulling doped single crystals, impurities in the melt flow faster at the vortex, resulting in lower resistivity [25] values in the middle of the two vortices.

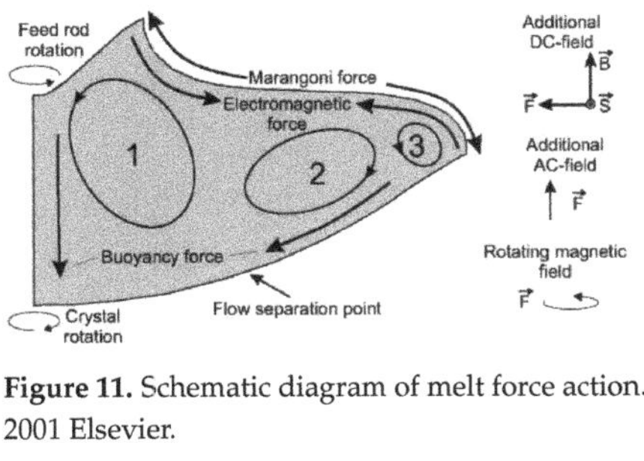

Figure 11. Schematic diagram of melt force action. Reprinted with permission from [22]; Copyright 2001 Elsevier.

4. Proof of Simulation Results

4.1. 2D Model Simulation Results

The simulation results of the 2D model, referring to the Refs. [26,27] were calculated for the melt temperature and flow under stepped and unstepped inductors. Figure 12 shows the melt temperature distribution in the absence and with flow, respectively. Without flow, the inductor steps reduce the temperature at the center of the melt and increase the temperature at the outer diameter of the inductor. To compare the effect of stepped and unstepped inductors more significantly on the melt, it is clear that the temperature in the center of the melt is lower with the addition of flow because the inductor step enhances melt flow.

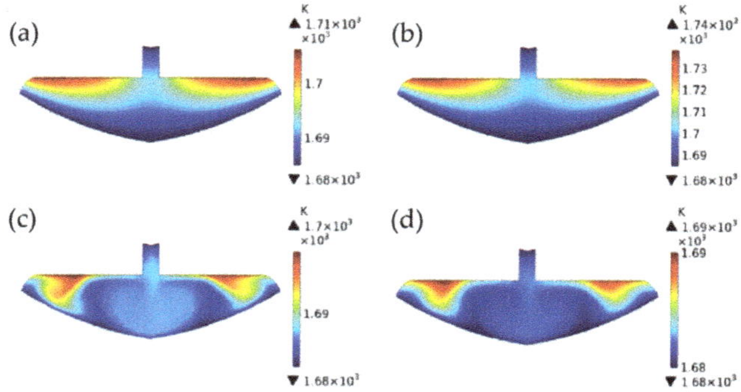

Figure 12. Temperature distribution in the melt without stepped inductor (**a**,**c**) and with stepped inductor (**b**,**d**) (**a**,**b**: without flow; **c**,**d**: with flow).

The increase of steps decreases the gradient of the distribution of the magnetic induction intensity, which makes the difference of the magnetic induction intensity within the polysilicon shrink. By comparing Figure 12c,d, it can be found that the maximum value of the melting zone temperature under the coils with steps is smaller. We have found that the resistivity distribution of the single crystals grown in the single-crystal coils with steps is more even when we conduct single-crystal growth experiments with the two kinds of coils, which is very useful for improving the quality of the single crystals.

To investigate the formation of vortices and their distribution in the melt due to the force, we chose the model with a stepped inductor underneath for convective calculations, time 10 min, frequency 2.5 MHz. We found three vortices in the flow results, as shown in Figure 13, formed by the combined action of EM, buoyancy, and Marangoni forces [28], which are also highly consistent with the results of force distribution in Figure 11. The flow is stronger at the center of the melt and second at the two ends. The reason is that the Marangoni effect dominates at the center and produces the largest vortex [29], while the two ends are close to the inductor, where the magnetic induction strength is high, and the vortex is formed mainly by EM forces. The flow results explain well the difference in temperature distribution in the melt above, and the 2D simulation results prove that the inductor steps greatly influence the thermal field and crystal growth in the melt.

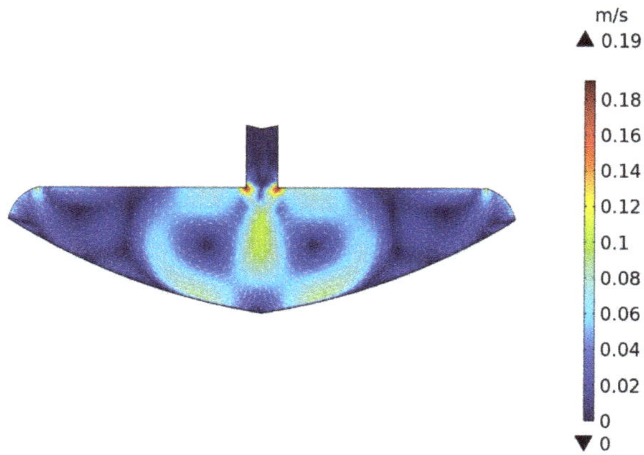

Figure 13. Distribution of the melt flow.

4.2. 3D Model Simulation Results

The 3D results refer to Ref. [30]; Figure 14 shows the current density distribution of two kinds of 3D inductors. The inductor slit changes can directly affect the distribution of the current and the induced magnetic field [31]. It can be seen that the current distribution of the one-way inductor is concentrated in the center around the hole and the slit, the current distribution of the outer side of the current distribution is less; it is because the density of the ring-effect currents is greatest at the inner diameter of the coil, the electromagnetic field generated is most concentrated at the "waist" of the melting zone of the crystals, thus generating the most Joule heat close to the center of the melting surface and the least Joule heat close to the outer surface of the polycrystalline silicon melt, which results in a very uneven distribution of radial temperatures across the entire polycrystalline silicon melt. Thus, the radial temperature distribution across the melting surface of polycrystalline silicon [32] is very uneven. However, the cross-way inductor, due to the existence of more slits, increased current from the main slit into the current shunt so that the current distribution is more uniform. At the same time, the current distribution in the outer side of the current distribution compared to the one-way inductor, the distribution of the current can be a greater degree of the effect of the melt so that the melt of the temperature is more uniformly distributed. The detailed calculation of temperature is mentioned in the following section.

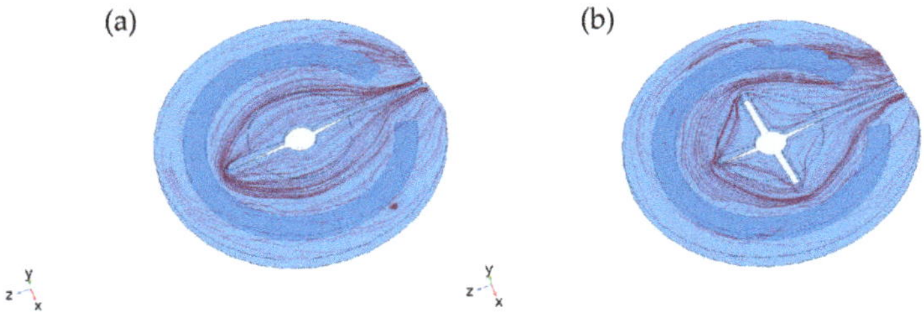

Figure 14. One-way Inductor (**a**) and cross-way Inductor (**b**) current density distribution.

In the following, the effect of the change in inductor structure on the melt temperature distribution. We calculated the temperature distribution of the melt transient for three different cases under two inductors separately. Figure 15 shows the melt temperature distribution for both inductors in the no flow, with flow and flow plus rotation cases, respectively. In the case of no flow, the temperature in both melting zones is mainly concentrated in the main silt, with little transfer to the secondary silt. When flow is added, the temperature distribution becomes relatively uniform and disperses the temperature of the main silt. The current is too concentrated since there is only one slit in the one-way inductor. A large amount of heat is distributed in the central hole and secondary silt during the FZ process. After adding rotation, the temperature is concentrated around the central hole, making the local temperature too high and the radial temperature gradient in the melt too large. To observe the radial temperature of the melt more intuitively, we intercept the radial width of the melt for temperature calculation, as shown in Figure 16, and get that the temperature gradient of the melt under the one-way inductor is larger than that under the cross-way coil, and it is most apparent near the center hole. In Figure 17, we calculated the axial temperature of the waist, measured at a distance of the line connecting the centers of the top and bottom of the "waist". Figures 17 and 18 show that the high-temperature zone in the "waist" of the one-way inductor is greater, and the axial temperature gradient is large, which will make the "waist" of the melt become very thin or even fracture in actual production.

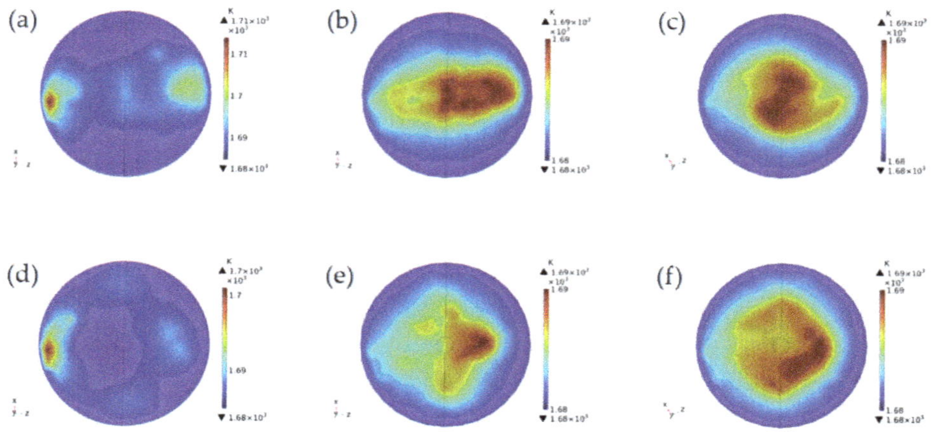

Figure 15. Temperature distribution in the melt under one-way inductor (**a**–**c**) and cross-way inductor (**d**–**f**) (**a**,**d**: without flow; **b**,**e**: with flow; **c**,**f**: flow with rotation).

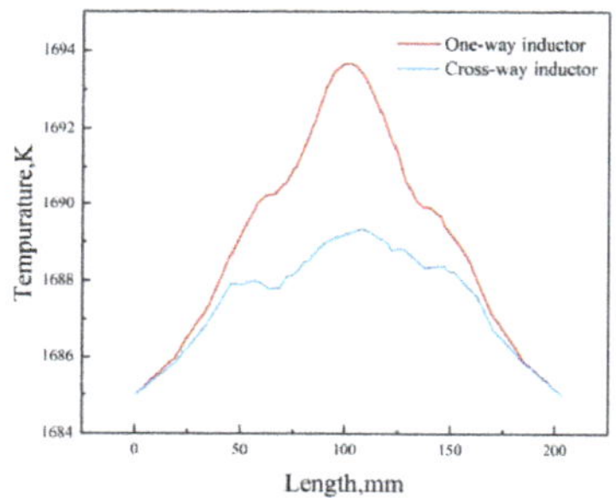

Figure 16. Comparison of radial temperature in the melt with rotation.

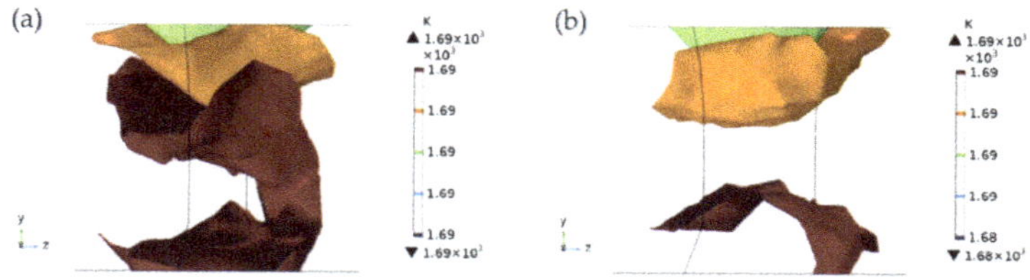

Figure 17. Equivalent surface distribution of "waist" temperature under one-way inductor (**a**) and cross-way inductor (**b**) with rotation.

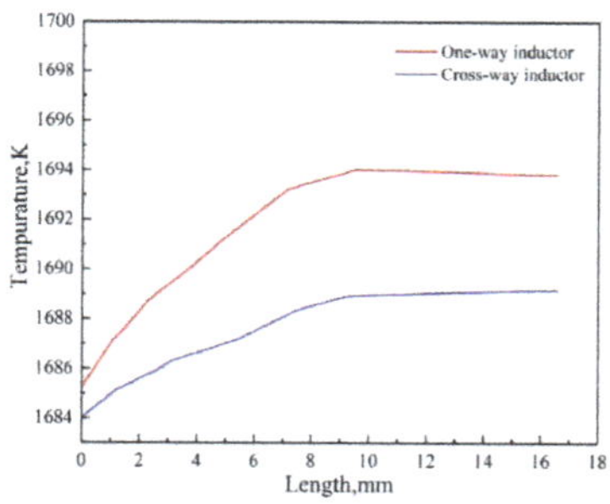

Figure 18. Comparison of axial temperature at the waist with rotation.

In contrast, the cross-way inductor has more slits and a large flux density distribution range, which makes the temperature distribution more uniform. There is no heat concentration near the center hole. Therefore, the slit of the inductor directly affects the temperature distribution of the melt and affects the crystal growth.

Figure 19 shows the flow distribution in the melt under both inductors, with the same scale factor for the velocity arrows. It can be found that under the cross inductor, the flow is stronger in all but the outermost part of the melt, the convection in both melts becomes uniform with rotation, and the convection in the melt under the cross-way inductor has a greater maximum value of velocity at the center due to the addition of more slits, which better solves the problem of thermal concentration. The 3D simulation results prove that the inductor slit greatly influences the temperature and flow in the melt, fundamentally due to the different inductor slit designs resulting in different distribution of magnetic induction lines. Based on this, it is proved that the design of the cross-way slit is more advantageous by calculating the thermal and flow fields.

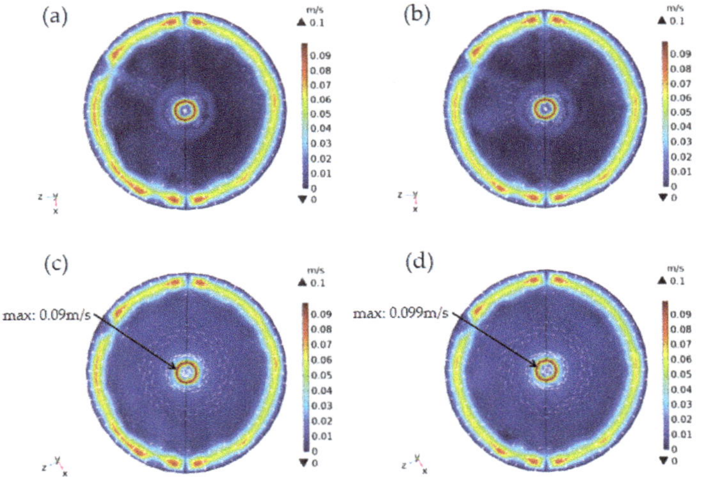

Figure 19. Flow distribution in the melt under one-way inductor (**a**,**c**) and cross-way inductor (**b**,**d**) (**a**,**b**: without rotation; **c**,**d**: with rotation).

5. Conclusions

The implementation of the FZ process proves that COMSOL can realize the calculation of complex models, covers a wide range of physics, and can realize the coupling calculation of multiple physical fields. This simulation analyzes the thermal and flow fields from two dimensions: 2D steps and 3D slits, and the results are consistent with the actual production situation. The calculation results show that the inductor structure can change the distribution of the EM field, temperature, and melt flow. The temperature and flow in the melt are more uniformly distributed with step inductors and cross-way inductors. The addition of steps decreases the center temperature of the melt zone, and the cross-way slit disperses the temperature of the center hole so that the temperature will not be overly concentrated in the "waist", and the thermal field becomes more stable.

Meanwhile, after calculating the simulation results, we refer to previous literature studies [33] and compare them to find that under certain growth parameter conditions, the addition of step and cross-way slits does enable the dislocation-free growth of FZ single crystal silicon. Therefore, the study of the effect of the change of growth parameters on the growth of FZ silicon is a problem that needs to be solved in the follow-up work. Meanwhile, the shortcoming of this study is that the modeling process did not include the "Deformation Geometry" module and only considered the influence of the inductor structure on the physical field of the melting zone, thus ignoring the change in the shape of the interface.

Based on the above, the step and cross slits design should be applied to inductors in actual production, making the FZ process more stable and improving the crystal quality. The direct simulation results obtained through numerical modeling were in good agreement with the actual production test. More importantly, using this numerical simulation method can reduce the amount of experimental and deepen the understanding of the FZ silicon single crystal growth system.

Author Contributions: Conceptualization, G.C.; Methodology, G.C.; Software, J.S.; Data curation, G.C.; Writing—original draft, X.A.; Supervision, C.S., H.Z., J.S., L.X., G.L. and G.C. All authors have read and agreed to the published version of the manuscript.

Funding: This work was Funded by the project for the Science and Technology Correspondent of Tianjin City (Grant No. 20YDTPJC01710) and the Research Foundation of Education Bureau of Hebei (Grant No. QN2021044).

Institutional Review Board Statement: Not applicable.

Informed Consent Statement: Not applicable.

Data Availability Statement: Not applicable.

Conflicts of Interest: The authors declare no conflict of interest.

References

1. Mohamed Ariff, A.H.; Hashmi, M.S.J.; Brabazon, D. Monocrystalline Silicon Grown Using Floating Zone Technique. In *Reference Module in Materials Science and Materials Engineering*; Elsevier: Amsterdam, The Netherlands, 2018; ISBN 978-0-12-803581-8.
2. Sueoka, K.; Kamiyama, E.; Vanhellemont, J. Theoretical Study of the Impact of Stress on the Behavior of Intrinsic Point Defects in Large-Diameter Defect-Free Si Crystals. *J. Cryst. Growth* **2013**, *363*, 97–104. [CrossRef]
3. Ratnieks, G.; Muižnieks, A.; Mühlbauer, A. Modelling of Phase Boundaries for Large Industrial FZ Silicon Crystal Growth with the Needle-Eye Technique. *J. Cryst. Growth* **2003**, *255*, 227–240. [CrossRef]
4. Feigelson, R.S. Crystal Growth History: Theory and Melt Growth Processes. *J. Cryst. Growth* **2022**, *594*, 126800. [CrossRef]
5. Taishi, T.; Ohno, Y.; Yonenaga, I. Constitutional Supercooling in Heavily As-Doped Czochralski Si Crystal Growth. *J. Cryst. Growth* **2014**, *393*, 42–44. [CrossRef]
6. Hossain, M.M.; Watauchi, S.; Nagao, M.; Tanaka, I. Effects of Growth Parameters on Silicon Molten Zone Formed by Infrared Convergent-Heating Floating Zone Method. *J. Cryst. Growth* **2017**, *459*, 105–111. [CrossRef]
7. Muižnieks, A.; Raming, G.; Mühlbauer, A.; Virbulis, J.; Hanna, B.; Ammon, W.V. Stress-Induced Dislocation Generation in Large FZ-and CZ-Silicon Single Crystals—Numerical Model and Qualitative Considerations. *J. Cryst. Growth* **2001**, *230*, 305–313. [CrossRef]
8. Dold, P. Analysis of Microsegregation in RF-Heated Float Zone Growth of Silicon—Comparison to the Radiation-Heated Process. *J. Cryst. Growth* **2004**, *261*, 1–10. [CrossRef]
9. Rost, H.-J.; Menzel, R.; Luedge, A.; Riemann, H. Float-Zone Silicon Crystal Growth at Reduced RF Frequencies. *J. Cryst. Growth* **2012**, *360*, 43–46. [CrossRef]
10. Rost, H.-J.; Buchovska, I.; Dadzis, K.; Juda, U.; Renner, M.; Menzel, R. Thermally Stimulated Dislocation Generation in Silicon Crystals Grown by the Float-Zone Method. *J. Cryst. Growth* **2020**, *552*, 125842. [CrossRef]
11. Wünscher, M.; Menzel, R.; Riemann, H.; Lüdge, A. Combined 3D and 2.5 D Modeling of the Floating Zone Process with Comsol Multiphysics. *J. Cryst. Growth* **2014**, *385*, 100–105. [CrossRef]
12. Rudevičs, A.; Muižnieks, A.; Ratnieks, G.; Mühlbauer, A.; Wetzel, T. Numerical Study of Transient Behaviour of Molten Zone during Industrial FZ Process for Large Silicon Crystal Growth. *J. Cryst. Growth* **2004**, *266*, 54–59. [CrossRef]
13. Muiznieks, A.; Virbulis, J.; Lüdge, A.; Riemann, H.; Werner, N. Floating Zone Growth of Silicon. In *Handbook of Crystal Growth*; Elsevier: Amsterdam, The Netherlands, 2015; pp. 241–279. ISBN 978-0-444-63303-3.
14. Yu, F.; Chen, W.; Zhang, D. The Research of Simulation on Eddy Current Separation Process Based on MATLAB and COMSOL. *Procedia CIRP* **2016**, *56*, 520–523. [CrossRef]
15. Han, X.-F.; Kakimoto, K.; Alradi, S.; Zaidat, K. Numerical Investigation of Floating Zone Silicon Using Halbach Array Magnets. *J. Cryst. Growth* **2020**, *546*, 125773. [CrossRef]
16. Menzel, R.; Rost, H.-J.; Kießling, F.M.; Sylla, L. Float-Zone Growth of Silicon Crystals Using Large-Area Seeding. *J. Cryst. Growth* **2019**, *515*, 32–36. [CrossRef]
17. Han, X.-F.; Liu, X.; Nakano, S.; Harada, H.; Miyamura, Y.; Kakimoto, K. 3D Numerical Simulation of Free Surface Shape during the Crystal Growth of Floating Zone (FZ) Silicon. *J. Cryst. Growth* **2018**, *483*, 269–274. [CrossRef]

18. Mühlbauer, A.; Muiznieks, A.; Virbulis, J. Analysis of the Dopant Segregation Effects at the Floating Zone Growth of Large Silicon Crystals. *J. Cryst. Growth* **1997**, *180*, 372–380. [CrossRef]
19. Mühlbauer, A.; Muiznieks, A.; Raming, G.; Ratnieks, G. 2D and 3D Numerical Modelling of FZ Large Silicon Single Crystal Growth: Relation between Growth Process and Crystal Quality. In Proceedings of the International Scientific Colloquium Modelling for Saving Resources, Riga, Latvia, 17–18 May 2001; pp. 17–22.
20. Nacke, B.; Muiznieks, A. Numerical Modelling of the Industrial Silicon Single Crystal Growth Processes. *GAMM-Mitteilungen* **2007**, *30*, 113–124. [CrossRef]
21. Ratnieks, G.; Muiznieks, A.; Buligins, L.; Raming, G.; Mühlbauer, A.; Lüdge, A.; Riemann, H. Influence of the Three Dimensionality of the HF Electromagnetic Field on Resistivity Variations in Si Single Crystals during FZ Growth. *J. Cryst. Growth* **2000**, *216*, 204–219. [CrossRef]
22. Han, X.-F.; Liu, X.; Nakano, S.; Kakimoto, K. Numerical Analysis of Dopant Concentration in 200 Mm (8 Inch) Floating Zone Silicon. *J. Cryst. Growth* **2020**, *545*, 125752. [CrossRef]
23. Raming, G.; Muižnieks, A.; Mühlbauer, A. Numerical Investigation of the Influence of EM-Fields on Fluid Motion and Resistivity Distribution during Floating-Zone Growth of Large Silicon Single Crystals. *J. Cryst. Growth* **2001**, *230*, 108–117. [CrossRef]
24. Lan, C.W.; Chian, J.H. Three-Dimensional Simulation of Marangoni Flow and Interfaces in Floating-Zone Silicon Crystal Growth. *J. Cryst. Growth* **2001**, *230*, 172–180. [CrossRef]
25. Lei, A.; Græsvænge, M.; Hindrichsen, C. Reduced Radial Resistivity Variation of FZ Si Wafers with Advanced NTD. *J. Cryst. Growth* **2019**, *512*, 65–68. [CrossRef]
26. Surovovs, K.; Muiznieks, A.; Sabanskis, A.; Virbulis, J. Hydrodynamical Aspects of the Floating Zone Silicon Crystal Growth Process. *J. Cryst. Growth* **2014**, *401*, 120–123. [CrossRef]
27. Sabanskis, A.; Virbulis, J. Simulation of the Influence of Gas Flow on Melt Convection and Phase Boundaries in FZ Silicon Single Crystal Growth. *J. Cryst. Growth* **2015**, *417*, 51–57. [CrossRef]
28. Jin, C.; Okano, Y.; Minakuchi, H.; Dost, S. Numerical Simulation of Thermo-Solutal Marangoni Convection in a Floating Half-Zone with Radiation Effects under Zero Gravity. *Int. J. Heat Mass Transf.* **2022**, *194*, 123010. [CrossRef]
29. Munakata, T.; Someya, S.; Tanasawa, I. Suppression of Marangoni Convection in the FZ Melt by High-Frequency Magnetic Field. *J. Cryst. Growth* **2002**, *235*, 167–172. [CrossRef]
30. Han, X.-F.; Liu, X.; Nakano, S.; Harada, H.; Miyamura, Y.; Kakimoto, K. 3D Numerical Study of the Asymmetric Phenomenon in 200 Mm Floating Zone Silicon Crystal Growth. *J. Cryst. Growth* **2020**, *532*, 125403. [CrossRef]
31. Plāte, M.; Dadzis, K.; Krauze, A.; Menzel, R.; Virbulis, J. Validation of a 3D Mathematical Model for Feed Rod Melting during Floating Zone Si Crystal Growth. *J. Cryst. Growth* **2019**, *521*, 46–49. [CrossRef]
32. Freiheit, H.-C.; Bonauer-Klepp, K.; Baumann, R. A General Method for the Calculation of Segregation Profiles in Floating Zone Grown Silicon Ingots with Non-Uniform Initial Distribution of the Solute. *J. Cryst. Growth* **2007**, *304*, 317–323. [CrossRef]
33. Menzel, R. Growth Conditions for Large Diameter FZ Si Single Crystals. Available online: https://api-depositonce.tu-berlin.de/server/api/core/bitstreams/1fc3bc7e-6cf3-4b97-b216-18d45ab871f0/content (accessed on 11 October 2013).

coatings

MDPI

Article

Controllable Low-Bias Rectifying Behaviors Induced by AA-P_2 Dopants in Armchair Silicene Nanoribbons with Different Widths

Caiping Cheng [1,2], Haibo Zhang [2,*], Haifeng Zhang [1] and Hui Yao [1]

[1] College of Basic Science, Shanxi Agricultural University, Taigu 030801, China
[2] College of Resources and Environment, Shanxi Agricultural University, Taigu 030801, China
* Correspondence: haiboz666@163.com; Tel.: +86-0354-6288322

Abstract: The electronic transport properties and rectifying behaviors of armchair silicene nanoribbons (ASiNRs) were investigated by using first-principles density functional theory, in which the left lead was pristine ASiNR and the right lead was doped ASiNR where two phosphorus (P) atoms replaced a pair of adjacent silicon atoms in the same sublattice A (AA-P_2). Two types of AA-P_2-doped models were considered for P dopant-substitute silicon atoms at the center or edges. The results showed that the rectification behavior of the system with a large rectifying ratio could be found, which could be modulated by changing the width of the silicene nanoribbons or the position of the AA-P_2 dopant. Mechanisms were revealed to explain the rectifying behaviors and provide a theoretical basis for semiconductor rectifier devices.

Keywords: silicene nanoribbon; dopants; electron transport; rectification; DFT

1. Introduction

The development of molecular electronics began in 1974, when Aviram and Ratner put forward the theory of ac molecular rectifier [1]. The brilliance of this work was that even a single molecule could exhibit superior electronic properties if it was properly designed as a silicon-based device. Since then, people have devoted themselves to the development of self-assembled single molecule devices and various experimental tools as well as the study of their charge transport characteristics in molecules [2–4]. Up to now, molecular electronics was a frontier science, which not only provided the possibility for the study of materials at the molecular level, but also met the technical requirements for the development of the miniaturization of traditional silicon-based electronic devices [5–7].

With the development trend of the miniaturization of integrated circuits, single-molecule functional devices seemed to be an effective solution to this problem. In recent decades, research on molecular switches, rectifiers, and transistors has been widely used [8–11]. As is known, molecular switches were the basic components of most electronic devices, and were widely used in information storage, logical data processing, signal processing, etc. Inspired by molecular rectifiers, more researchers became devoted to the exploration and reporting of various molecular rectifiers. However, early research on molecular-rectifying devices was limited by the materials used. Stokbro et al., theoretically demonstrated a molecular rectification with negligible rectification [12]. Ford et al., and Stadler et al. reported that rectifying behaviors with a very small rectifying ratio were found in their devices [13,14]. In conclusion, it seemed difficult to achieve ideal large rectifying molecular devices in previous studies. Therefore, with the discovery of new materials, it is necessary to study the rectifiers of other materials with a high-performance molecular rectifier.

Since the introduction of graphene [15], researchers have studied hundreds of two-dimensional materials due to their novel electronic properties, but only a few have been theoretically defined as Dirac materials. A theoretical study showed that monolayer silicene was a stable Dirac material with a curved honeycomb structure and that it had the same order of magnitude of carrier mobility as graphene. Meanwhile, the electronic properties of silicene were more controllable than graphene. Due to its excellent characteristics and intrinsic connection with traditional silicon-based nanotechnology, silicene has become the most promising alternative material for future electronic devices [16–21]. Theoretical calculations have shown that silicene is a type of semi-metallic nanomaterial with zero band gaps at the Fermi level. However, the successful preparation of silicene nanoribbons (SiNRs) experimentally solved this problem. Concurrently, a theoretical simulation showed that SiNRs had two types of structures, zigzag-edged SiNRs (ZSiNRs) and armchair-edged SiNRs (ASiNRs). Among them, ZSiNRs always exhibited metallic properties whereas ASiNRs could exhibit both metallic and semiconductor properties due to the different nanoribbon widths. ASiNRs whose edges were passivated by hydrogen atoms had a width-dependent energy gap and ferromagnetism was observed in ZSiNRs passivated along their edges. The electronic properties of SiNRs can be changed by external electric fields [22], doping [23], passivation [24], chemical functionalization [25], defects [26], etc. Among them, impurity atom doping is the most direct and effective method.

In SiNRs, the elements aluminum (Al), phosphorus (P), and sulfur (S) can replace the silicon atoms, similar to B and N elements [27–32]. Sivek et al., studied the effect of the doping of B, N, Al, and P on the electronic structure and magnetism of SiNRs, showing that the doping of B, N, and P atoms provided electron acceptors and Al atoms acted as electron donors [33]. In addition, the study of the electronic structure of B/N co-doped SiNRs by Zhang et al. showed that the band gap of ASiNRs could be effectively regulated by adjusting the site of the B/N co-doping [34,35]. Moreover, a previous work only investigated the effect of AA-P_2 doping on the electronic properties of ASiNRs with a width of 7 at the same site [36]. However, a systematic study on the electron transport performance of asymmetric electrode rectification devices based on AA-P_2-doped ASiNRs with different widths has not been carried out. Thus, in this paper, we systematically studied the influence of ASiNRs with different widths on the electronic properties and constructed electronic devices with AA-P_2-doped asymmetric electrodes. The research results showed that there was an obvious negative differential resistance effect and strong rectification behavior. The results could provide a theoretical basis for the research of molecular devices.

2. Computational Method and Model

In this paper, we calculated the electronic structure of AA-P_2-doped ASiNRs with two quasi-adjacent substitutional phosphorus atoms incorporated in pairs of neighboring silicon atoms in the same sublattice A by first-principles based on density functional theory. After the test, the electronic stability model of the AA-P_2-doped ASiNR system was determined only; the metallic properties of zigzag SiNRs (ZSiNRs) could not be modulated by the location of AA-P_2. On this basis, a two-probe transport model of ASiNRs diagonally doped with AA-P_2 as an asymmetric electrode was constructed, as shown in Figure 1a. It was composed of three parts, including a left electrode, a center scattering area, and a right electrode. The electrode was composed of two repeated silicon cells and the central scattering region was composed of four repeated silicon cells. It is well-known that SiNRs are similar to graphene nanoribbons (GNRs) in that their intrinsic properties depend on the width (W) of the nanoribbon; these were divided into three categories according to the width of the nanoribbon (W = 3n, 3n + 1, and 3n + 2, where n was an integer). Therefore, we selected ASiNRs with three widths—N = 6, 7, and 8, respectively—and these were named A6, A7, and A8. In addition, the center (edge) of the right electrode was diagonally embedded into the P atom pair and the constructed two-probe model was defined as A6/7/8-C/E (indicating ASiNRs with widths of 6, 7, and 8 as center-doped or edge-doped with AA-P_2, as shown in Figure 1b–d.

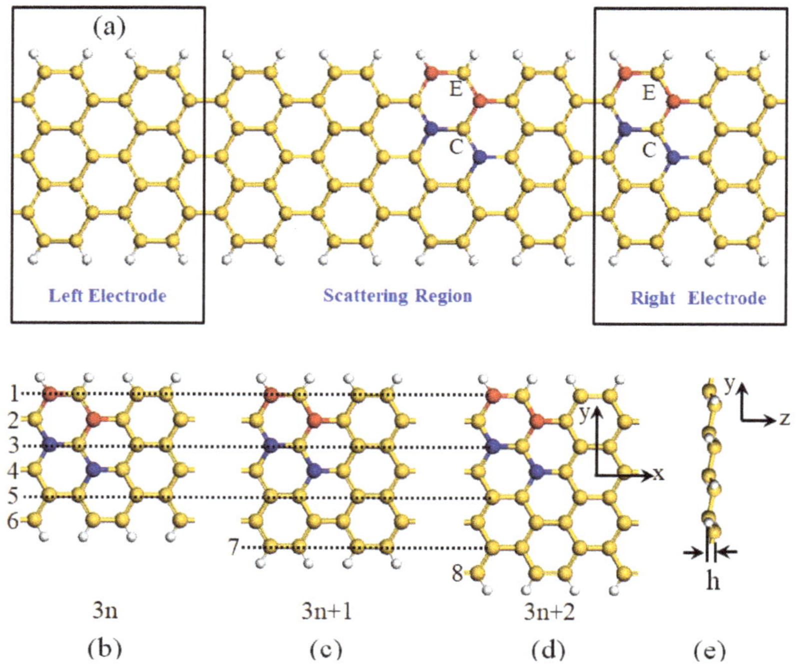

Figure 1. (**a**) The schematic illustration of two-probe system based on ASiNR device. (**b–d**) Model of AA$-P_2\angle$ASiNRs with different widths: N = 3n (6); 3n + 1 (7); and 3n + 2 (8). The colored atoms represent substitutional doping of AA$-P_2$ and the labels with C and E denote the dopants at the center (blue atoms) and the edge (red atoms), respectively. (**e**) Side view of ASiNRs. The white in each diagram represents hydrogen atoms.

Using a first-principles calculation based on density functional theory, the geometry and electronic band structure of the system were optimized by the SIESTA software package [37]. The form of Perdew, Burke, and Ernzerhof uses the generalized gradient approximation (GGA) to represent the exchange correlation potential and the basic set of atomic orbitals uses the double ξ polarization (DZP) function [38,39]. The maximum cut-off kinetic energy was set to 200 Ry. For the N-ASiNRs system, we considered periodic boundary conditions in the x-direction of the coordinates and the distance between the two cells in the Z-direction was greater than 10 Å to ensure the isolated monolayer structure of the system. The Brillouin zone (BZ) was sampled using 15 × 1 × 1 Monkhorst–Pack k-points in the self-consistent total energy calculations. The TRANSIESTA-C software package was used to calculate the transport performance of the tow probes [40]. The voltage transmission spectrum was calculated according to the standard equation:

$$T(E,V) = Tr[\Gamma_L(E,V)\, G(E,V)\, \Gamma_R(E,V)\, G^{\dagger}(E,V)] \quad (1)$$

where G is the Green's function of the contact region, $\Gamma_{L/R}$ is the coupling matrix, and V is the applied voltage bias. The computational formula for a current could be used; namely, the Landauer–Büttiker formula:

$$I = 2e/h \int_{-\infty}^{+\infty} dE[f_L(E-\mu_L) - f_R(E-\mu_R)]T(E,V) \quad (2)$$

where μ_L and μ_R are the electrochemical potentials for the two electrodes and T(E, V) is the transmission coefficient at energy E and bias voltage V.

3. Results and Discussion

As the stable structure of AA-P_2-diagonally doped ASiNRs (AA-P_2∠ASiNRs) had been determined in previous studies [31], we gave the band structure of ideal central-doped and edge-doped ASiNRs, as shown in Figure 2a–c. The ideal ASiNRs with widths of 3n and 3n + 1 showed semiconductivity and the ASiNRs with a width of 3n + 2 showed typical metal properties. However, both central and edge doping produce two impurity bands near the Fermi level. It is well-known that the doping band affects the electron transfer near the Fermi level, so we constructed a double probe transport model with the doping system as the central scattering region. On this basis, we first explored the current voltage (I–V) characteristics of the AA-P_2-diagonally doped ASiNR device; the voltage range was selected as −1.0 V to 1.0 V with a step of 0.05 V, as shown in Figure 3a–c. It was evident that the I–V curves of almost all devices were asymmetric and exhibited typical semiconductor characteristics at corresponding threshold voltages. Compared with center doping, the edge doping system was more obvious. For example, when the voltage increased from 0 V to 0.6 V, the current of the A7-C system increased, but when the voltage was greater than 0.6 V, the current value of the system decreased instead; thus, a negative differential resistance effect appeared. The same was true for A6-C. According to the traditional definition of the rectification coefficient by $R(V) = I(+V)/I(-V)$, the current values of the positive and negative voltages were in the same order of magnitude. When the current under a positive bias was much larger than the corresponding current under a negative bias, the rectification behavior was stronger. As can be seen from Figure 4, within the range of 0.1 V–0.4 V, the rectification coefficient of A7-E decreased with the increase in the threshold voltage; it reached a maximum of 190 at 0.3 V, which was several times greater than the rectification ratio achieved by the molecular devices previously reported in the literature [32,41,42] (Table 1). In addition, compared with the doping of a single P atom reported by the research group of Zhang et al. [43], the doping of AA-P_2 could generate more impurity energy bands, resulting in a more obvious negative differential effect and rectification behavior of rectifier devices under asymmetric electrodes. The research of nanodevices provides a theoretical basis. Other rectification coefficients vary in the range of 0–10. The rectification ratio of A7-C was smaller than that of A7-E and the variation range was not large; it was less than 20 at almost all threshold voltages. Meanwhile, the A8-E system achieved a maximum rectification ratio of 35 at a bias of 0.1 V. However, for the AA-P_2∠ASiNR device with a width of 6, the I–V curve was obviously different from that with a width of 7, showing a high symmetry. Therefore, a large proportion of rectification could only be obtained in the A7-E system; it was completely negligible in the other AA-P_2∠ASiNR systems.

Table 1. Comparison of the maximum rectification ratio of different non-metals.

Molecular Device	Rectification Ratio	References
Al/P co-doped silicene nanoribbon systems	88	Kharwar et al. (2022) [32]
B/N-doped graphene nanoribbons	18	Pramanik et al. (2012) [41]
B/N top angle-doped with regular triangle graphene sheets	99	Ling et al. (2015) [42]
AA-P2-doped silicene nanoribbon devices	190	This study

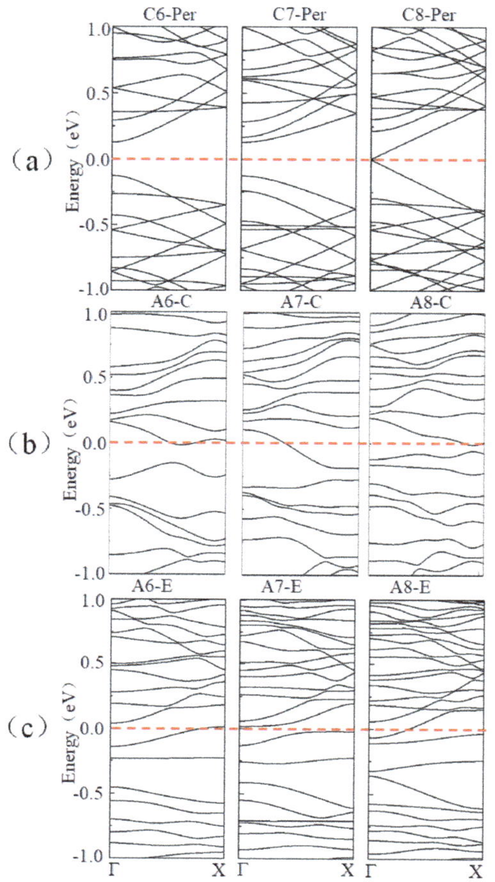

Figure 2. (**a**–**c**) The electronic band structures of original ASiNRs, AA-P_2-central-doped ASiNRs, and AA-P_2-edge-doped ASiNRs with widths of 6, 7, and 8 (N = 6, 7, and 8), respectively. The red dotted line is the Fermi energy level.

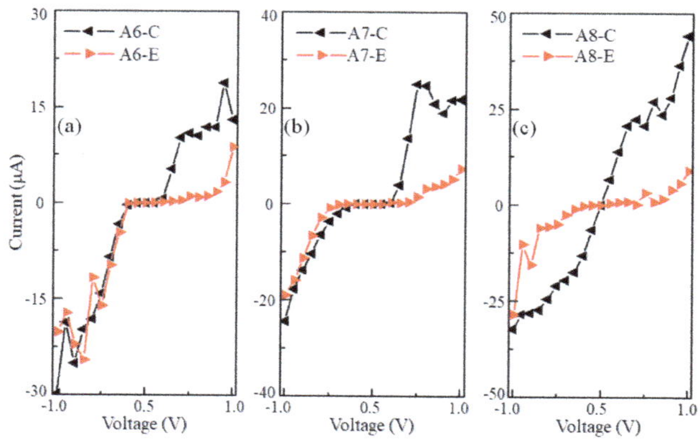

Figure 3. The I−V characteristics for AA−P^2∠ASiNR devices with widths of 6, 7, and 8 (N = 6, 7, and 8) in (**a**−**c**), respectively.

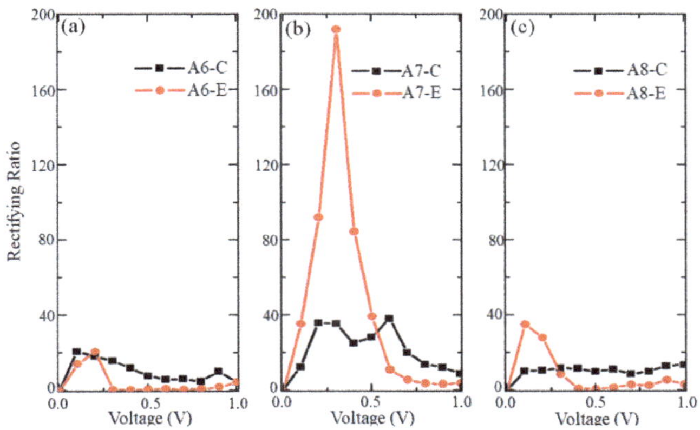

Figure 4. The computed rectification ratio for the considered ASiNRs: (**a**) ASiNRs with a width of 6 (N = 6); (**b**) ASiNRs with a width of 7 (N = 7); (**c**) ASiNRs with a width of 8 (N = 8), respectively.

In order to understand the strong rectification effect, we calculated the transmission coefficient spectra of AA-P_2∠ASiNRs with widths of 6, 7, and 8 at 0 V, 0.5 V, and −0.5 V biases, respectively, as shown in Figure 5. It is well-known that the transmission coefficient spectrum and the integral area in the bias window determine the magnitude of the current. First, the three central- or edge-doped AA-P_2 devices with different widths exhibited different family behaviors at a zero bias. When an additional bias was applied, we observed a more interesting phenomenon: for the A7-E system, a strong and wide peak in the transmission spectrum appeared in the bias window of V = 0.5 V, resulting in a large current in the device. However, when the bias V = −0.5 V, almost no peak could be found in the displayed bias window, so the system had a strong rectification behavior. For the A7-C system, although the peak value of the transmission spectrum was obvious in the V = 0.5 V bias window, when the opposite bias V = −0.5 V was applied, the transmission peak value was almost unchanged. This indicated that the magnitude of current under a positive and negative bias of the A7-C device was similar and the rectification effect was not obvious. When we changed the width of AA-P_2∠ASiNR to 8, the bias window of the A8-C device appeared with several peaks at 0.5 V with a certain current. At the bias of −0.5 V, the transmission peak was very small, indicating that the rectification ratio of A8-C had a certain fluctuation. For the A8-E system, a sharp peak appeared near the Fermi level at the 0.5 V bias and the peak almost disappeared at the −0.5 V bias. Therefore, the device also exhibited weak rectification behavior. However, for the AA-P_2∠ASiNR device with a width of 6, a strong and wide transmission peak appeared at the positive and negative 0.5 V bias. Due to the strong metallic properties, rectification behavior did not occur in the A6 device. In short, the rectification effect of the edge-doped AA-P_2∠ASiNRs was more superior.

In order to clearly elucidate the rectification phenomenon of the AA-P_2∠ASiNR devices, the band structures of the left and right electrodes and corresponding transmission spectra of A6-E, A7-E, and A8-E at 0 V, 0.5 V, and −0.5 V, respectively, were further presented, as shown in Figure 6. In this part, we only describe the edge-doped system (E) because almost all the central-doped system (C) exhibited metal characteristics and the rectification coefficient was relatively small compared with the E system. It is well-known that the magnitude of a current depends not only on the good matching of the band structures corresponding with the left and right electrodes, but also on the delocalization of the molecular projected self-consistent Hamiltonian (MPSH) energy state at the corresponding eigenvalue (E_V). Therefore, we also calculated the MPSH energy states of the A6-E, A7-E, and A8-E devices under certain E_Vs, as shown in Figure 7. As shown in Figure 5, it was found that although the ideal ASiNRs where N = 6 and 7 exhibited semiconductor properties and the ideal ASiNRs where N = 8 exhibited gold properties, all the ASiNRs doped with AA-P_2 exhibited different band structures. The results showed that two new bands appeared near the Fermi level of the doped electrode (right electrode) and the impurity band changed with the width of the ASiNRs. It is well-known that each phosphorus atom can act as an electron donor and generate a new band around the Fermi energy according to the rules of electron arrangement outside the nucleus of phosphorus atoms, which is when impurity phosphorus atoms were introduced into the ideal ASiNRs. By doping the P atom pair, a delocalized state of the system was generated in the conduction band (CB) rather than in the valence band (VB). Thus, the extra electrons introduced by the impurity atom pair caused the valence band of the right electrode to move all the way down compared with the pristine ASiNRs. When a positive voltage was applied to the device, the energy band of the left electrode moved down and the energy band of the right electrode moved up compared with the Fermi level. With the movement of the energy bands of the left and right electrodes, the VB of the left electrode contacted the CB of the right electrode, forming an electronic channel, and then the transmission spectrum appeared in the transmission spectrum.

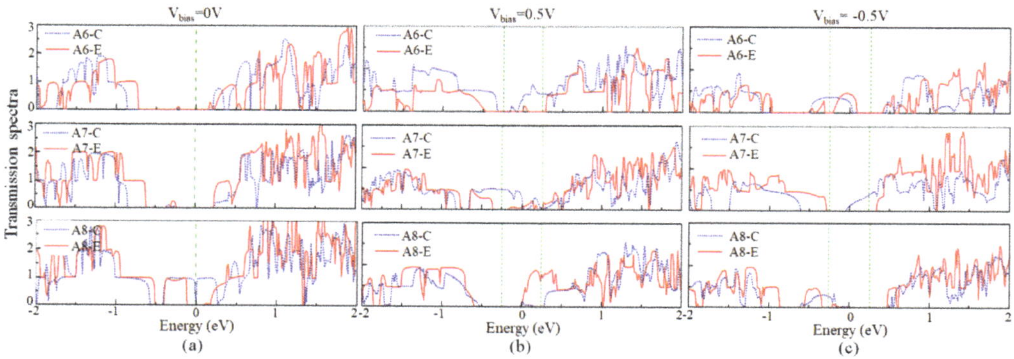

Figure 5. The transmission spectra of AA-P_2-doped ASiNR devices with widths of 6, 7, and 8 under different biases (N = 6, 7, and 8); **(a–c)** correspond with V_b = 0 V, 0.5 V, and −0.5 V, respectively. The vertical green dashed lines indicate the bias window.

Due to the semiconductor characteristics of the ideal and AA-P_2-doped ASiNRs with a width of 6, an energy gap of about 0.52 eV appeared in the A6-E device near the Fermi level at the zero bias, as shown in Figure 6a. When a bias of 0.5 V was applied to A6-E, the transmission peak did not appear in the bias window, which depended on the fact that the VB of the left electrode near the Fermi level did not match the CB of the right electrode. However, when the same negative bias was applied, the band of the left electrode shifted up and the band of the right electrode shifted down and a transmission peak could be observed near the Fermi level, depending on the maximum VB of the left electrode matching the impurity band of the right electrode, as shown in Figure 6c. At the same time, the MPSH electron state at a −0.5V bias of the A6-E system at E_v = −0.21 eV was completely localized on the right side and the localization on the left side of the scattering region disappeared, which showed that there was a weak coupling between the scattering area and the ideal ASiNRs of the left electrode and signified that the transmission channel in the A6-E system under the −0.5V bias was suppressed, as shown in Figure 7a. It could be seen that there were few states available for electron movement through the A6-E device, making the transmission coefficient relatively small; thus, a large proportion of rectification behavior could not be obtained.

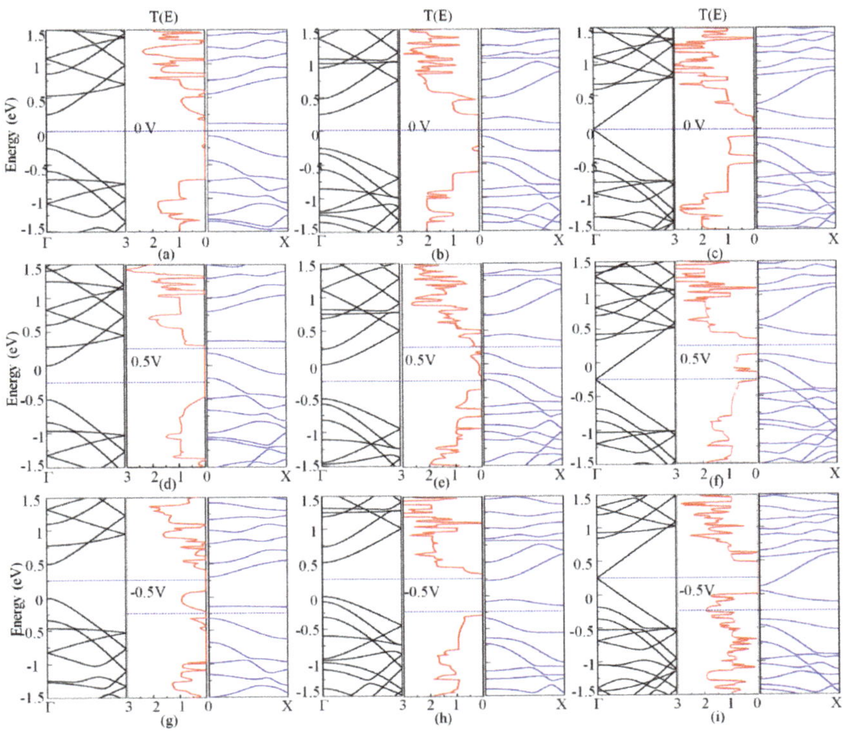

Figure 6. Band structure for the left lead (left panels), transmission curve (middle panels), and band structure for the right lead (right panels). (**a–c**) A6/7/8-E devices under a zero bias (0 V); (def) A6/7/8-E devices under a 0.5 V bias; (**g–i**) A6/7/8-E devices under a −0.5 V bias. The blue dashed lines correspond with the chemical potentials of the left and right electrodes.

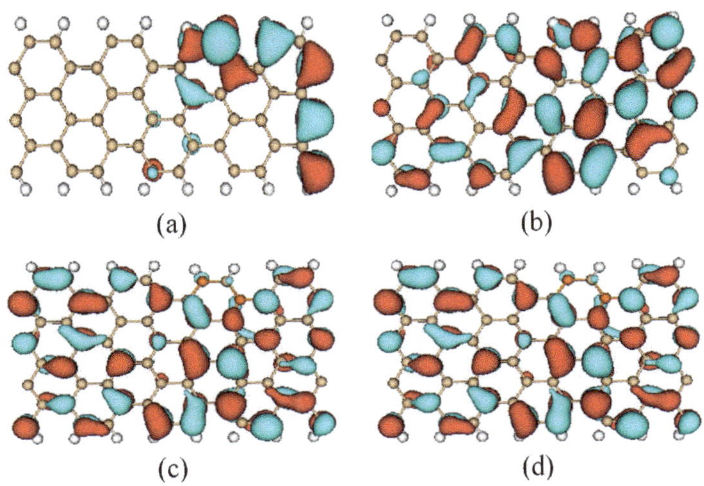

Figure 7. (**a**) MPSH of A6-E at an eigenvalue of $E_v = -0.21$ eV at a bias of 0.5 V; (**b**) MPSH of A7-E at an eigenvalue of $E_v = 0.15$ eV at a bias of 0.5 V; (**c,d**) MPSH of A8-E at an eigenvalue of $E_v = -0.07$ eV at a bias of 0.5 V and −0.5 V, respectively.

As shown in Figure 6b, the impurity bands in the A7 system with edge-doped AA-P_2 were distributed around the Fermi level and the energy gap was reduced compared with the pristine system. Under a zero bias, the upper part of the left electrode VB and the lower part of the right electrode CB had no matching energy band, so there was a relatively large gap between the valence band and the impurity band in the transmission spectrum, which was in the range of −0.25 eV–0.25 eV. When a positive bias was applied, a sharp peak appeared within the bias window, which varied greatly from the zero bias. At the same time, as shown in Figure 7b, the MPSH of A7-E at E_v = 0.15 eV had a good delocalization state and formed an effective electron transport channel. However, when a negative bias was applied, the maximum VB of the left electrode did not match the impurity band of the right electrode in the bias window and there was no transmission peak. Therefore, the total amplitude of the transmission coefficient under a positive bias (0.5 V) was much larger than that under a negative bias (−0.5 V), resulting in a strong rectification effect.

For the A8-E device, the band mechanism of the perfect ASiNRs system exhibited typical metallic properties, the energy gap was opened by doping AA-P_2, and the band structure was changed, as shown in Figure 6c. At a zero bias, the maximum value of the valence band of the left electrode did not match the minimum value of the conduction band of the right electrode in the energy range of −0.85 eV to 0.25 eV. Therefore, there was no transmission peak near the Fermi level. However, a strong transmission coefficient could clearly be seen in the rest of the energy range. When a 0.5 V bias was applied, the valence band part of the left electrode matched well with the conduction band of the right electrode due to the effect of the impurity band of the right electrode. As a result, a strong transmission coefficient appeared in the bias window. When a bias voltage of −0.5 V was applied, the conduction band of the left electrode partially matched the valence band of the right electrode; there was also a strong transmission coefficient within the bias window. The above results showed that the electron localization of MPSH at E_v = −0.07 eV under biases of 0.5 V and −0.5 V was good and the energy state distribution was very similar, as shown in Figure 7c,d. Therefore, there was no obvious rectification behavior.

According to the above analysis, ASiNRs with a width W = 3n + 2 exhibited metallic properties whereas for W = 3n or 3n + 1, the system exhibited semiconductor properties and there was a certain energy gap near the Fermi level. When AA-P_2 was doped into the semiconductor ASiNRs with widths of W = 3n or 3n + 1, the extra electron of the P atom was one more than that of the C atom; thus, these extra electrons filled the original empty conduction band and a corresponding doping band appeared near the Fermi level. Interestingly, the doping of AA-P_2 still left an energy gap between the valence band and the conduction band of the ASiNR system. However, for metallic ASiNRs with the width W = 3n + 2, the energy gap opened near the Fermi level after the impurity band of AA-P_2 doping, showing the semiconductor characteristics.

We now summarize the rectifying mechanism of the AA-P_2-doped semiconductor ASiNR heterojunction. At a zero bias, although an impurity band appeared near the Fermi level on the AA-P_2-doped right electrode due to the band gap on the left electrode, the transmittance remained zero. When a positive bias was applied, the band structure of the left electrode shifted upward whereas the band structure of the right electrode shifted downward. When the lowest conduction band of the left electrode and the highest conduction band of the right electrode were in contact with each other, electrons could be conducted from one electrode to the other, so an effective current could be observed. However, when a negative bias was applied, the band of the left electrode shifted downward whereas the right electrode shifted upward. Although the conduction band of the left electrode moved to the Fermi level, the energy band gap of the right electrode also moved to the Fermi level, so the electrons could not be conducted from the left electrode to the right electrode; thus, the transmission coefficient of the bias window was still zero and no current was formed. In this way, the rectifying behaviors of positive and negative voltages were more obvious.

Obviously, in the system exhibiting rectifying behaviors, there was a switching voltage during the current increase from zero, which depended on the energy gap of the left electrode. Therefore, as the width of the nanoribbon increased, the band gap correspondingly decreased, making the switching voltage in the I–V curve shift to the left. However, for systems with the widths 3n or 3n + 1, there was still a band gap near the Fermi level due to the effect of the impurity band after AA-P_2 doping. In other words, all ASiNRs had a partially filled conduction band with a gap between the conduction band and the valence band, so a clear rectified behavior could be observed. As the width of the nanoribbon increased, the band gap of the electrode decreased, making the rectification region as a whole narrower and moving toward zero. In this way, it was deduced that when the bandwidth was wide enough, the rectification behavior would disappear.

4. Conclusions

In summary, the influence of nanoribbon-width-dependent AA-P_2 doping on the electron transport performance of ASiNRs was investigated by using density functional theory based on first-principles. The results showed that the ASiNR devices exhibited different I–V characteristics with the width of the nanoribbon and the AA-P_2 doping position. Compared with center doping, edge doping had a more obvious influence on the I–V characteristic curve of the device, resulting in a serious asymmetry of the current on a positive and negative bias. Especially for the A7-E system, the obvious asymmetry of the I–V curve indicated that the system had a strong rectification behavior, which made the rectification rate close to 200. However, for other ASiNR devices, the rectification coefficient was completely negligible. A further investigation showed that the strong rectifying behaviors of the A7-E device depended on a good matching between the VBs of the left electrode and the CBs of the right electrode. Meanwhile, the MPSH of A7-E at $E_v = 0.15$ eV had a good delocalization state and formed an effective electron transport channel. However, under the same negative bias, the valence band of the left electrode and the conduction band of the right electrode were completely separated near the Fermi level, which made the positive and negative bias have a very strong rectification behavior. These results are expected to provide a theoretical reference value for the research and development of silicon-based nanodevices.

Author Contributions: Conceptualization, C.C. and H.Z. (Haibo Zhang); methodology, C.C. and H.Z. (Haifeng Zhang); software, C.C. and H.Z. (Haifeng Zhang); validation, H.Z. (Haifeng Zhang) and H.Y.; investigation, C.C. and H.Y.; writing—review and editing, C.C. and H.Z. (Haibo Zhang); funding acquisition, C.C. and H.Z. (Haibo Zhang) All authors have read and agreed to the published version of the manuscript.

Funding: This research was financially supported by Scientific and Technological Innovation Programs of Higher Education Institutions in Shanxi (2020L0162), the Science and Technology Innovation Fund of Shanxi Agricultural University (2015YJ20), the National Youth Fund Project (52100149), and the Applied Basic Research Program of Shanxi Province (202103021223169).

Institutional Review Board Statement: Not applicable.

Informed Consent Statement: Not applicable.

Data Availability Statement: Data sharing is not applicable to this article.

Conflicts of Interest: The authors declare no conflict of interest.

References

1. Aviram, A.; Ratner, M.A. Molecular rectifiers. *Chem. Phys. Lett.* **1974**, *29*, 277. [CrossRef]
2. Love, J.C.; Estroff, L.A.; Kriebel, J.K.; Nuzzo, R.G.; Whitesides, G.M. Self-Assembled Monolayers of Thiolates on Metals as a Form of Nanotechnology. *Chem. Rev.* **2005**, *105*, 1103–1169. [CrossRef] [PubMed]
3. Jeong, H.; Kim, D.; Xiang, D.; Lee, T. High-Yield Functional Molecular Electronic Devices. *ACS Nano* **2017**, *11*, 6511–6548. [CrossRef] [PubMed]

4. Choi, E.; Jang, Y.; Ho, D.; Chae, W.; Earmme, T.; Kim, C.; Seo, S. Development of Dithieno [3, 2-b: 2′, 3′-d] thiophene (DT Derivatives as Solution-Processable Small Molecular Semiconductors for Organic Thin Film Transistors. *Coatings* **2021**, *11*, 122 [CrossRef]
5. Xiang, D.; Wang, X.; Jia, C.; Lee, T.; Guo, X. Molecular-Scale Electronics: From Concept to Function. *Chem. Rev.* **2016**, *11* 4318–4440. [CrossRef]
6. Cui, Y.; Xia, C.; Su, Y.; Zhang, B.; Zhang, T.; Liu, Y.; Hu, Z.; Tang, X. Switching characteristics of anthraquinone molecular device based on graphene electrodes. *Acta Phys. Sin.* **2021**, *70*, 038501. [CrossRef]
7. Zhang, X.-H.; Li, J.-Z.; Qin, F.; Wang, L. Experiments and numerical simulations of microelectronic devices subjected to multipoin impact loading. *Microelectro. Reliabili.* **2021**, *127*, 114408. [CrossRef]
8. Welker, M.E. Ferrocenes as Building Blocks in Molecular Rectifiers and Diodes. *Molecules* **2018**, *23*, 1551. [CrossRef]
9. Abbas, M.A.; Hanoon, F.H.; Al-Badry, L.F. Possibility designing half-wave and full-wave molecular rectifiers by using singl benzene molecule. *Phys. Lett. A* **2018**, *382*, 608–612. [CrossRef]
10. Thompson, D.; Del Barco, E.; Nijhuis, C.A. Design principles of dual-functional molecular switches in solid-state tunnel junction *Appl. Phys. Lett.* **2020**, *117*, 030502. [CrossRef]
11. Baidya, A.; Lenka, T.R.; Baishya, S. 3D Double-Gate Junctionless Nanowire Transistor-Based Pass Transistor Logic Circuits fo Digital Applications. *IETE J. Res.* **2019**, *68*, 1342–1349. [CrossRef]
12. Stokbro, K.; Taylor, J.; Brandbyge, M. Do Aviram-Ratner diodes rectify. *J. Am. Chem. Soc.* **2003**, *125*, 3674–3675. [CrossRef [PubMed]
13. Ford, M.; Hoft, R.C.; McDonagh, A.M.; Cortie, M.B. Rectification in donor-acceptor molecular junctions. *J. Phys. Condens. Matte* **2008**, *20*, 374106. [CrossRef] [PubMed]
14. Stadler, R.; Geskin, V.; Cornil, J. A theoretical view of unimolecular rectification. *J. Phys. Condens. Matter.* **2008**, *20*, 37410 [CrossRef] [PubMed]
15. Novoselov, K.S.; Geim, A.K.; Morozov, S.V.; Jiang, D.; Zhang, Y.; Dubonos, S.V.; Grigorieva, I.V.; Firsov, A.A. Electric Field Effec in Atomically Thin Carbon Films. *Science* **2004**, *306*, 666. [CrossRef] [PubMed]
16. Wu, Z.; Zhang, Y.; Li, G.; Du, S.; Gao, H. Electronic properties of silicene in BN/silicene van der Waals heterostructures. *Chin Phys. B* **2018**, *27*, 077302. [CrossRef]
17. Grazianetti, C.; Cinquanta, E.; Molle, A. Two-dimensional silicon: The advent of silicone. *2D Mater.* **2016**, *3*, 012001. [CrossRef]
18. Drummond, N.; Zolyomi, V.; Falko, V. Electrically tunable band gap in silicone. *Phys. Rev. B* **2012**, *85*, 075423. [CrossRef]
19. Ezawa, M. Valley-Polarized Metals and Quantum Anomalous Hall Effect in Silicene. *Phys. Rev. Lett.* **2012**, *109*, 055502. [CrossRef
20. Liu, G.; Lei, X.L.; Wu, M.S.; Xu, B.; Ouyang, C.Y. Comparison of the stability of free-standing silicene and hydrogenated silicen in oxygen: A first principles investigation. *J. Phys. Conden. Matt.* **2014**, *26*, 355007. [CrossRef]
21. Cesano, F.; Scarano, D. Graphene and Other 2D Layered Hybrid Nanomaterial-Based Films: Synthesis, Properties, and Applica tions. *Coatings* **2018**, *12*, 419. [CrossRef]
22. Tsai, W.-F.; Huang, C.-Y.; Chang, T.-R.; Lin, H.; Jeng, H.-T.; Bansil, A. Gated silicene as a tunable source of nearly 100% spin-polarized electrons. *Nat. Commun.* **2013**, *4*, 1500. [CrossRef] [PubMed]
23. Lopez, A. Substitutional doping widens silicene gap. *J. Phys. Chem. C* **2014**, *118*, 18788–18792. [CrossRef]
24. Zhang, D.; Long, M.; Xie, F.; Ouyang, J.; Xu, H.; Gao, Y. Hydrogenations and electric field induced magnetic behaviors in armchai silicene nanoribbons. *Sci. Rep.* **2016**, *6*, 23677. [CrossRef]
25. Gao, N.; Zheng, W.T.; Jiang, Q. Density functional theory calculations for two-dimensional silicene with halogen functionalization *Phys. Chem. Chem. Phys.* **2012**, *14*, 257–261. [CrossRef] [PubMed]
26. Iordanidou, K.; Houssa, M.; Broek, B.V.D.; Pourtois, G.; Afanas'Ev, V.V.; Stesmans, A. Impact of point defects on the electroni and transport properties of silicene nanoribbons. *J. Condens. Matter. Phys.* **2016**, *28*, 035302. [CrossRef] [PubMed]
27. Jiang, Q.G.; Zhang, J.F.; Ao, Z.M.; Huang, H.J.; Wu, Y.P. Tuneable electronic and magnetic properties of hybrid silicone/silican nanoribbons induced by nitrogen doping. *Thin Solid Film.* **2018**, *653*, 126–135. [CrossRef]
28. Sarebanha, B.; Ahmadi, S. Doping Effect on Spin Dependent Electron Transport in Zigzag Silicene Nanoribbons. *Procedia Mater Sci.* **2015**, *11*, 259–264. [CrossRef]
29. Sahin, H.; Peeters, F.M. Adsorption of alkali, alkaline-earth, and 3d transition metal atoms on silicone. *Phys. Rev. B* **2013** *87*, 085423. [CrossRef]
30. Lin, X.; Ni, J. Much stronger binding of metal adatoms to silicene than to graphene: A first-principles study. *Phys. Rev. B* **2012** *86*, 075440. [CrossRef]
31. Cocoletzi, H.H.; Aguila, J.E.C. DFT studies on the Al, B, and P doping of silicone. *Sup. Mic.* **2018**, *114*, 242–250. [CrossRef]
32. Kharwarl, S.; Singh, S. First-Principles Investigations of Al/P Co-doping Effect on Disconnected Armchair Silicene Nanoribbon for Switching Device Applications. *Silicon* **2022**, *14*, 8609–8619. [CrossRef]
33. Sivek, J.; Sahin, H.; Partoens, B.; Peeters, F.M. Adsorption and absorption of boron, nitrogen, aluminum, and phosphorus on silicone: Stability and electronic and phonon properties. *Phys. Rev. B* **2013**, *87*, 085444. [CrossRef]
34. Zheng, F.-B.; Zhang, C.-W.; Yan, S.-S.; Li, F. Novel electronic and magnetic properties in N or B doped silicene nanoribbons *J. Mater. Chem. C* **2013**, *1*, 2735–2743. [CrossRef]
35. Ma, L.; Zhang, J.-M.; Xu, K.-W.; Ji, V. Structural and electronic properties of substitutionally doped armchair silicene nanoribbons *Phys. B Condens. Matt.* **2013**, *425*, 66–71. [CrossRef]

36. Cheng, C.; Hu, H.; Zhang, Z.; Zhang, H. Perfect rectifying behavior induced by AA-P_2 dopants in armchair silicene nanoribbon devices. *RSC Adv.* **2016**, *6*, 7042. [CrossRef]
37. Perdew, J.P.; Burke, K.; Ernzerhof, M. Generalized gradient approximation made simple. *Phys. Rev. Lett.* **1996**, *77*, 3865. [CrossRef]
38. Brandbyge, M.; Mozos, J.-L.; Ordejon, P.; Taylor, J.; Stokbro, K. Density-functional method for nonequilibrium electron transport. *Phys. Rev. B* **2002**, *65*, 165401. [CrossRef]
39. Soler, J.; Artacho, E.; Gale, J.; Garcia, A.; Junquera, J.; Ordejon, P.; Sanchez-Portal, D. The SIESTA method for ab initio order-N materials simulation. *J. Phys. Condens. Matter.* **2002**, *14*, 2745. [CrossRef]
40. Taylor, J.; Guo, H.; Wang, J. Ab initio modeling of quantum transport properties of molecular electronic devices. *Phys. Rev. B* **2001**, *63*, 245407. [CrossRef]
41. Pramanik, A.; Sarkar, S.; Sarkar, P. Doped GNR p-n Junction as High Performance NDR and Rectifying Device. *J. Phys. Chem. C* **2012**, *116*, 18064–18069. [CrossRef]
42. Ling, Y.; Ning, F.; Zhou, Y.; Chen, K. Rectifying behavior and negative differential resistance in triangular graphene p–n junctions induced by vertex B-N mixture doping. *Org. Electron.* **2015**, *19*, 92–97. [CrossRef]
43. Zhou, Y.; Zhang, D.; Zhang, J.; Ye, C.; Miao, X. Negative differential resistance behavior in phosphorus-doped armchair graphene nanoribbon junctions. *J. Appl. Phys.* **2014**, *115*, 073703. [CrossRef]

MDPI AG
Grosspeteranlage 5
4052 Basel
Switzerland
Tel.: +41 61 683 77 34

Coatings Editorial Office
E-mail: coatings@mdpi.com
www.mdpi.com/journal/coatings

www.ingramcontent.com/pod-product-compliance
Lightning Source LLC
Chambersburg PA
CBHW062059220926
51850CB00010B/2825